The Little Book of Big Primes

Paulo Ribenboim

The Little Book of Big Primes

Springer-Verlag
New York Berlin Heidelberg London
Paris Tokyo Hong Kong Barcelona

Paulo Ribenboim
Department of Mathematics and Statistics
Queen's University
Kingston, Ontario
K7L 3N6 Canada

Mathematics Subject Classification: 11A41, 11B39, 11A51

Library of Congress Cataloging-in-Publication Data
Ribenboim, Paulo.
    The little book of big primes / Paulo Ribenboim.
        p.   cm.
    Includes bibliographical references and index.
    ISBN 0-387-97508-X (New York : acid free paper). — ISBN
3-540-97508-X (Berlin : acid free paper)
        1. Numbers, Prime   I. Title.
    QA246.R472   1991
    512'72—dc20   90-25061

Printed on acid-free paper.

Photocomposed copy prepared using LaTeX.
Printed and bound by R.R. Donnelley & Sons, Inc., Harrisonburg, VA.
Printed in the United States of America.

9 8 7 6 5 4 3 2 1

ISBN 0-387-97508-X Springer-Verlag New York Berlin Heidelberg
ISBN 3-540-97508-X Springer-Verlag Berlin Heidelberg New York

Nel mezzo del cammin di nostra vita
mi ritrovai per una selva oscura
che la diritta via era smarrita

Dante Alighieri, *L'Inferno*

Preface

This book could have been called "Selections from the Book of Prime Number Records." However, I prefered the title which propelled you on the first place to open it, and perhaps (so I hope) to buy it!

Richard K. Guy, with his winning ways, suggested the title to me, and I am grateful.

But the book isn't very different from its parent. Like a bonsai, which has all the main characteristics of the full-sized tree, this little paperback should exert the same fatal attraction. I wish it to be as dangerous as the other one, in the saying of John Brillhart. I wish that you, young student, teacher or retired mathematician, engineer, computer buff, all of you who are friends of numbers, to be driven into thinking about the beautiful theory of prime numbers, with its inherent mystery. I wish you to exercise your brain and fingers—not vice-versa.

But I do not wish you, specialist in number theory to look at this little book—most likely you have been eliminated from this shorter version—what a terrible feeling. But do not cry, you had your book already. This one is for those who will be taking over and should put their steps forward, mostly little, occasionally giant, to develop the science of numbers.

Paulo Ribenboim

Contents

5 Which Special Kinds of Primes Have Been Considered?   161

6 Heuristic and Probabilistic Results about Prime Numbers   179

Guiding the Reader

If a notation, which is not self-explanatory, appears without explanation on, say page 103, look at the Index of Notations, which is organized by page number; the definition of the notation should appear before page 103.

If you wish to see where and how often your name is quoted in this book, turn to the Index of Names, at the end of the book. Should I say that there is no direct relation between achievement and number of quotes earned?

If, finally, you do not want to read the book but you just want to have some information about Cullen numbers—which is perfectly legitimate, if not laudable—go quickly to the Subject Index. Do not look under the heading "Numbers," but rather "Cullen." And for a subject like "Strong Lucas pseudoprimes," you have exactly three possibilities . . . .

Index of Notations

The following traditional notations are used in the text without explanation:

Notation	Explanation
$m \mid n$	the integer m divides the integer n
$m \nmid n$	the integer m does not divide the integer n
$p^e \| n$	p is a prime, $p^e \mid n$ but $p^{e+1} \nmid n$
$\gcd(m, n)$	greatest common divisor of the integers m, n
$\ell\mathrm{cm}(m, n)$	least common multiple of the integers m, n
$\log x$	natural logarithm of the real number $x > 0$
$\mathbf{Z}$	ring of integers
$\mathbf{Q}$	field of rational numbers
$\mathbb{R}$	field of real numbers
$\mathbf{C}$	field of complex numbers

The following notations are listed as they appear in the book:

Page	Notation	Explanation
3	p_n	the nth prime
3	$p\#$	product of all primes q, $q \leq p$
5	F_n	nth Fermat number, $F_n = 2^{2^n} + 1$
15	g_p	smallest primitive root modulo p
18	$[x]$	the largest integer in x, that is, the only integer such that $[x] \leq x < [x] + 1$
22	$\phi(n)$	totient or Euler's function
23	$\lambda(n)$	Carmichael's function
24	$\omega(n)$	number of distinct prime factors of n
25	$N_\phi(m)$	$\#\{\, n \geq 1 \mid \phi(n) = m \,\}$
27	t_n^*	primitive part of $a^n - b^n$
27	$k(m)$	square-free kernel of m
29	(a/p)	Legendre symbol
30	(a/b)	Jacobi symbol

Page	Notation	Explanation
89	$s\ell\mathrm{psp}(P,Q)$	strong Lucas pseudoprime with parameters (P,Q)
100	p_n	a prime number with n digits
105	$\pi(x)$	the number of primes p, $p \leq x$
108	$\mu(n)$	Möbius function
112	$\nu(f,N)$	$\#\{\,x \mid x = 0,1,\ldots,N \text{ such that } |f(x)| \text{ is equal to 1 or to a prime}\,\}$
120	$f(x) \sim h(x)$	f, h are asymptotically equal
120	$f(x) = g(x) + o(h(x))$	the difference $f(x) - g(x)$ is negligible in comparison to $h(x)$
121	$\zeta(s)$	Riemann's zeta function
122	B_k	Bernoulli number
123	$S_k(n)$	$= \sum_{j=1}^{n} j^k$
124	$B_k(X)$	Bernoulli polynomial
125	$\mathrm{Li}(x)$	logarithmic integral
126	$\theta(x)$	$= \sum_{p \leq x} \log p$, Tschebycheff function
127	$\mathrm{Re}(s)$	real part of s
127	$\Gamma(s)$	gamma function
127	γ	Euler's constant
128	$J(x)$	weighted prime-power counting function
129	$R(x)$	Riemann's function
130	$\Lambda(n)$	von Mangoldt's function
131	$\psi(x)$	summatory function of the von Mangoldt function
133	$M(x)$	Mertens' function
135	$\phi(x,m)$	$= \#\{a \mid 1 \geq a \geq x, a \text{ isnot a multiple of } 2,3,\ldots,P_m\}$
137	$N(T)$	$= \#\{\rho = \delta + it \mid 0 \geq \delta \geq 1, \zeta(\rho) = 0, 0 < t \geq T\}$
137	ρ_n	nth zero of $\zeta(s)$ in the upper half of the critical strip
142	d_n	$= p_{n+1} - p_n$
154	P_k	set of all k-almost-primes
146	$\pi_2(x)$	$= \#\{\,p \text{ prime} \mid p + 2 \text{ is also a prime and } p + 2 \leq x\,\}$
146	B	Brun's constant
147	C_2	$= \prod_{p>2}\left(1 - \frac{1}{(p-1)^2}\right)$, twin prime constant
148	$\pi_{2k}(x)$	$= \#\{\,n \geq 1 \mid p_n \leq x \text{ and } p_{n+1} - p_n = 2k\,\}$

xvi   Index of Notations

Introduction

The *Guinness Book of Records* became famous as an authoritative source of information to settle amiable disputes between drinkers of, it was hoped, the Guinness brand of stout. Its immense success in recording all sorts of exploits, anomalies, endurance performances, and so on has in turn influenced and sparked these very same performances. So one sees couples dancing for countless hours or persons buried in coffins with snakes, for days and days—just for the purpose of having their name in this bible of trivia. There are also records of athletic performances, extreme facts about human size, longevity, procreation, etc.

Little is found in the scientific domain. Yet, scientists—mathematicians in particular—also like to chat while sipping wine or drinking a beer in a bar. And when the spirits mount, bets may be exchanged about the latest advances, for example, about recent discoveries concerning numbers.

Frankly, if I were to read in the *Whig-Standard* that a brawl in one of our pubs began with a heated dispute concerning which is the largest known pair of twin prime numbers, I would find this highly civilized.

But not everybody agrees that fights between people are desirable, even for such all-important reasons. So, maybe I should reveal some of these records. Anyone who knows better should not hesitate to pass me additional information.

I'll restrict my discussion to prime numbers: these are natural numbers, like 2, 3, 5, 7, 11, ..., which are not multiples of any smaller natural number (except 1). If a natural number is neither 1 nor a prime, it is called a composite number.

Prime numbers are important, since a fundamental theorem in arithmetic states that every natural number greater than 1 is a product of prime numbers, and moreover, in a unique way.

Without further ado it is easy to answer the following question: "Which is the oddest prime number?" It is 2, because it is the only even prime number!

There will be plenty of opportunities to encounter other prime numbers, like 1093 and 608,981,813,029, possessing interesting distinctive properties. Prime numbers are like cousins, members of the same family, resembling one another, but not quite alike.

Facing the task of presenting the records on prime numbers, I was led to think how to organize this volume. In other words, to classify the main lines of investigation and development of the theory of prime numbers.

It is quite natural, when studying a set of numbers—in this case the set of prime numbers—to ask the following questions, which I phrase informally as follows:

How many? How to decide whether an arbitrary given number is in the set? How to describe them? What is the distribution of these numbers, both at large and in short intervals? And, then, to focus attention on distinguished types of such numbers, as well as to experiment with these numbers and make predictions—just as in any science.

Thus, I have divided the presentation into the following topics:

(1) How many prime numbers are there?

(2) How to recognize whether a natural number is a prime?

(3) Are there functions defining prime numbers?

(4) How are the prime numbers distributed?

(5) Which special kinds of primes have been considered?

(6) Heuristic and probabilistic results about prime numbers.

The discussion of these topics will lead me to indicate the relevant records.

1

How Many Prime Numbers Are There?

The answer is given by the fundamental theorem:

There exist infinitely many prime numbers.

I shall present five and two half(!) proofs of this theorem. Some suggest interesting developments; other are just clever or curious. There are of course more (but not quite infinitely many) proofs of the existence of infinitely many primes.

I. Euclid's Proof

Suppose that $p_1 = 2 < p_2 = 3 < \cdots < p_r$ are all the primes. Let $P = p_1 p_2 \cdots p_r + 1$ and let p be a prime dividing P; then p cannot be any of $p_1, p_2, \ldots, p_r$, otherwise p would divide the difference $P - p_1 p_2 \cdots p_r = 1$, which is impossible. So this prime p is still another prime, and $p_1, p_2, \ldots, p_r$ would not be all the primes. $\qquad\square$

I shall write the infinite increasing sequence of primes as

$$p_1 = 2, p_2 = 3, p_3 = 5, p_4 = 7, \ldots, p_n, \ldots.$$

Euclid's proof is pretty simple; however, it does not give any information about the new prime, only that it is at most equal to the number P, but it may well be smaller.

For every prime p, let $p\#$ denote the product of all primes q such that $q \leq p$. Following a suggestion of Dubner (1987), $p\#$ may be called the *primorial* of p.

The answer to the following questions are unknown:

Are there infinitely many primes p for which $p\# + 1$ is prime?

Are there infinitely many primes p for which $p\# + 1$ is composite?

RECORD

The only known primes p for which $p\#+1$ is a prime, are $p = 2, 3, 5,$ 7, 11, 31, 379, 1019, 1021, 2657, 3229, 4547, 4787, 11549, and 13649. The record is due to Dubner (1987). Previous work was done by Borning (1972), Templer (1980), Buhler, Crandall & Penk (1982). It is also known that $p\# + 1$ is composite for every other prime $p < 11213$.

Here is a variant of the above problem. Consider the sequence $q_1 = 2$, $q_2 = 3$, $q_3 = 7$, $q_4 = 43$, $q_5 = 139$, $q_6 = 50{,}207$, $q_7 = 340{,}999$, $q_8 = 2{,}365{,}347{,}734{,}339, \ldots$, where q_{n+1} is the highest prime factor of $q_1 q_2 \cdots q_n + 1$ (so $q_{n+1} \neq q_1, q_2, \ldots, q_n$). In 1963, Mullin asked: Does the sequence $(q_n)_{n \geq 1}$ contain all the prime numbers? Does it exclude at most finitely many primes? Is the sequence monotonic?

Cox & van der Poorten found in 1968 congruences which are sufficient to decide whether a given prime is excluded; moreover, they have shown that the primes 5, 11, 13, 17, 19, 23, 29, 31, 37, 41, and 47 do not appear in the sequence. They have also conjectured that there exist infinitely many excluded primes.

In 1985, Odoni considered the similar sequence:

$$w_1 = 2, \ w_2 = 3, \ldots, w_{n+1} = w_1 w_2 \ldots w_n + 1,$$

and he showed that there exist infinitely many primes which do not divide any number of the sequence, and of course, there exist infinitely many primes which divide some number of the sequence.

I $\frac{1}{2}$. Kummer's Proof

In 1878, the great Kummer gave the following variant of Euclid's proof—so I count it just as "half a proof."

Suppose that there exist only finitely many primes $p_1 < p_2 < \cdots < p_r$. Let $N = p_1 p_2 \cdots p_r > 2$. The integer $N - 1$, being a product of primes, has a prime divisor p_i in common with N; so, p_i divides $N - (N - 1) = 1$, which is absurd! □

This proof, by an eminent mathematician, is like a pearl, round, bright, and beautiful in its simplicity.

A very similar proof was given by another great mathematician, Stieltjes, in 1890.

II. Pólya's Proof

Pólya's proof (see Pólya & Szegö, 1924) uses the following idea: It is enough to find an infinite sequence of natural numbers $1 < a_1 < a_2 < a_3 < \cdots$ that are pairwise relatively prime (i.e., without common prime factor). So, if p_1 is a prime dividing a_1, if p_2 is a prime dividing a_2, etc., then $p_1, p_2, \ldots,$ are all different.

For this proof, the numbers a_n are chosen to be the Fermat numbers $F_n = 2^{2^n} + 1$ $(n \geq 0)$. Indeed, it is easy to see, by induction on m, that $F_m - 2 = F_0 F_1 \cdots F_{m-1}$; hence, if $n < m$, then F_n divides $F_m - 2$.

If a prime p would divide both F_n and F_m, then it would divide $F_m - 2$ and F_m, hence also 2, so $p = 2$. But F_n is odd, hence not divisible by 2. This shows that the Fermat numbers are pairwise relatively prime. □

Explicitly, the first Fermat numbers are $F_0 = 3$, $F_1 = 5$, $F_2 = 17$, $F_3 = 257$, $F_4 = 65537$, and it is easy to see that they are prime numbers. F_5 already has 10 digits, and each subsequent Fermat number is about the square of the preceding one, so the sequence grows very quickly. A natural task is to determine explicitly whether F_n is a prime, or at least to find a prime factor of it. I shall return to this point in Chapter 2.

It would be desirable to find other infinite sequences of pairwise relatively prime integers, without already assuming the existence of infinitely many primes. In a paper of 1964, Edwards examined this question and indicated various sequences, defined recursively, having this property. For example, if S_0, a are relative prime integers, with $S_0 > a \geq 1$, the sequence defined by the recursive relation

$$(\text{S1}) \qquad S_n - a = S_{n-1}(S_{n-1} - a) \quad (\text{for } n \geq 1)$$

consists of pairwise relatively prime natural numbers.

Similarly, if S_0 is odd and

$$(\text{S2}) \qquad S_n = S_{n-1}^2 - 2 \quad (\text{for } n \geq 1),$$

then, again, the integers S_n are pairwise relatively prime.

In the best situation, that is, when $S_0 = 3$, $a = 2$, the sequence (S1) is in fact the sequence of Fermat numbers: $S_n = F_n = 2^{2^n} + 1$.

The sequence (S2), which grows essentially just as quickly, has been considered by Lucas, and I shall return to it in Chapter 2.

In 1947, Bellman gave the following method to produce infinite sequences of pairwise relatively prime integers, without using the fact that there exist infinitely many primes. One begins with a non-constant polynomial $f(x)$, with integer coefficients, such that $f(0) \neq 0$, and such that if n, $f(0)$ are relatively prime integers, then $f(n)$, $f(0)$ are also relatively prime integers. Then, let $f_1(x) = f(x)$ and for $m \geq 1$, let $f_{m+1}(x) = f(f_m(x))$.

If it happens that $f_m(0) = f(0)$ for every $m \geq 1$ and if n, $f(0)$ are relatively prime, then the integers n, $f_1(n)$, $f_2(n)$, ..., $f_m(n)$, ... are pairwise relatively prime. For example, $f(x) = (x-1)^2 + 1$ satisfies the required conditions and, in fact, $f_n(-1) = 2^{2^n} + 1$—so, back to Fermat numbers!

II $\frac{1}{2}$. Schorn's Proof

The following variant of Pólya's proof was kindly communicated to me by P. Schorn and is unpublished elsewhere.

First observe that if $1 \leq i < j \leq n$ then $\gcd((n!)i+1, (n!)j+1) = 1$. Indeed, writing $j = i + d$ then $1 \leq d < n$, so

$$\gcd((n!)i + 1, (n!)j + 1) = \gcd((n!)i + 1, (n!)d) = 1,$$

because every prime p dividing $(n!)d$ is at most equal to n.

Now, if the number of primes would be m, taking $n = m + 1$, the above remark implies that the $m + 1$ integers $(m + 1)!i + 1$ $(1 \leq i \leq m + 1)$ are pairwise relatively prime, so there exist at least $m + 1$ distinct primes, contrary to the hypothesis. □

III. Euler's Proof

This is a rather indirect proof, which, in some sense is unnatural; but, on the other hand, as I shall indicate, it leads to the most important developments.

Euler showed that there must exist infinitely many primes because a certain expression formed with all the primes is infinite.

If p is any prime, then $1/p < 1$; hence, the sum of the geometric series is

$$\sum_{k=0}^{\infty} \frac{1}{p^k} = \frac{1}{1 - (1/p)}.$$

Similarly, if q is another prime, then

$$\sum_{k=0}^{\infty} \frac{1}{q^k} = \frac{1}{1 - (1/q)}.$$

Multiplying these equalities:

$$1 + \frac{1}{p} + \frac{1}{q} + \frac{1}{p^2} + \frac{1}{pq} + \frac{1}{q^2} + \cdots = \frac{1}{1 - (1/p)} \times \frac{1}{1 - (1/q)}.$$

Explicitly, the left-hand side is the sum of the inverses of all natural numbers of the form $p^h q^k$ ($h \geq 0$, $k \geq 0$), each counted only once, because every natural number has a unique factorization as a product of primes. This simple idea is the basis of the proof.

Euler's Proof. Suppose that $p_1, p_2, \ldots, p_n$ are all the primes. For each $i = 1, \ldots, n$

$$\sum_{k=0}^{\infty} \frac{1}{p_i^k} = \frac{1}{1 - (1/p_i)}.$$

Multiplying these n equalities, one obtains

$$\prod_{i=1}^{n} \left(\sum_{k=0}^{\infty} \frac{1}{p_i^k} \right) = \prod_{i=1}^{n} \frac{1}{1 - (1/p_i)},$$

and the left-hand side is the sum of the inverses of all natural numbers, each counted once—this follows from the fundamental theorem that every natural number is equal, in a unique way, to the product of primes.

But the series $\sum_{n=1}^{\infty}(1/n)$ is divergent; being a series of positive terms, the order of summation is irrelevant, so the left-hand side is infinite, while the right-hand side is clearly finite. This is absurd. □

In Chapter 4 I'll return to developments along this line.

IV. Washington's Proof

Washington's (1980) is a proof via commutative algebra. The ingredients are elementary facts of the theory of principal ideal domains, unique factorization domains, Dedekind domains, and algebraic numbers, and may be found in any textbook on the subject, such as Samuel's (1967) book: there is no mystery involved. First, I recall the needed facts:

1. In every number field (of finite degree) the ring of algebraic integers is a Dedekind domain: every nonzero ideal is, in a unique way, the product of prime ideals.

2. In every number field (of finite degree) there are only finitely many prime ideals that divide any given prime number p.

3. A Dedekind domain with only finitely many prime ideals is a principal ideal domain, and as such, every nonzero element is, up to units, the product of prime elements in a unique way.

Washington's Proof. Consider the field of all numbers of the form $a + b\sqrt{-5}$, where a, b are rational numbers. The ring of algebraic integers in this field consists of the numbers of the above form, with a, b ordinary integers. It is easy to see that 2, 3, $1 + \sqrt{-5}$, $1 - \sqrt{-5}$ are prime elements of this ring, since they cannot be decomposed into factors that are algebraic integers, unless one of the factors is a "unit" 1 or -1. Note also that

$$(1 + \sqrt{-5})(1 - \sqrt{-5}) = 2 \times 3,$$

the decomposition of 6 into a product of primes is not unique up to units, so this ring is not a unique factorization domain; hence, it is not a principal ideal domain. So, it must have infinitely many prime ideals (by fact 3 above) and (by fact 2 above) there exist infinitely many prime numbers. □

V. Fürstenberg's Proof

This is an ingenious proof based on topological ideas. Since it is so short, I cannot do any better than transcribe it verbatim; it appeared in 1955:

In this note we would like to offer an elementary "topological" proof of the infinitude of the prime numbers. We introduce a topology into the space of integers S, by using the arithmetic progressions (from $-\infty$ to $+\infty$) as a basis. It is not difficult to verify that this actually yields a topological space. In fact, under this topology, S may be shown to be normal and hence metrizable. Each arithmetic progression is closed as well as open, since its complement is the union of the other arithmetic progressions (having the same difference). As a result, the union of any finite number of arithmetic progessions is closed. Consider now the set $A = \cup A_p$, where A_p consists of all multiples of p, and p runs through the set of primes ≥ 2. The only numbers not belonging to A are -1 and 1, and since the set $\{-1, 1\}$ is clearly not an open set, A cannot be closed. Hence A is not a finite union of closed sets which proves that there are an infinity of primes.   □

Golomb developed further the idea of Fürstenberg, and wrote an interesting short paper in 1959.

2

How to Recognize Whether a Natural Number is a Prime

In the art. 329 of *Disquisitiones Arithmeticae*, Gauss (1801) wrote:

> The problem of distinguishing prime numbers from composite numbers and of resolving the latter into their prime factors is known to be one of the most important and useful in arithmetic... The dignity of the science itself seems to require that every possible means be explored for the solution of a problem so elegant and so celebrated.

The first observation concerning the problem of primality and factorization is clear: there is an algorithm for both problems. By this, I mean a procedure involving finitely many steps, which is applicable to every number N and which will indicate whether N is a prime, or, if N is composite, which are its prime factors. Namely, given the natural number N, try in succession every number $n = 2$, 3, ... up to $\lceil \sqrt{N} \rceil$ (the largest integer not greater than $\sqrt{N}$) to see whether it divides N. If none does, then N is a prime. If, say, N_0 divides N, write $N = N_0 N_1$, so $N_1 < N$, and then repeat the same procedure with N_0 and with N_1. Eventually this gives the complete factorization into prime factors.

What I have just said is so evident as to be irrelevant. It should, however, be noted that for large numbers N, it may take a long time with this algorithm to decide whether N is a prime or composite.

This touches the most important practical aspect, the need to find an efficient algorithm—one which involves as few operations as possible, and therefore requires less time and costs less money to be performed.

It is my intention to divide this chapter into several sections in which I will examine various approaches, as well as explain the required theoretical results.

I. The Sieve of Eratosthenes

As I have already said, it is possible to find if N is a prime using trial division by every number n such that $n^2 \leq N$.

Since multiplication is an easier operation than division, Eratosthenes (in the 3rd century BC) had the idea of organizing the computations in the form of the well-known sieve. It serves to determine all the prime numbers, as well as the factorizations of composite numbers, up to any given number N. This is illustrated now for $N = 101$.

Do as follows: write all the numbers up to 101; cross out all the multiples of 2, bigger than 2; in each subsequent step, cross out all the multiples of the smallest remaining number p, which are bigger than p. It suffices to do it for $p^2 < 101$.

	2	3	~~4~~	5	~~6~~	7	~~8~~	~~9~~	~~10~~
11	~~12~~	13	~~14~~	~~15~~	~~16~~	17	~~18~~	19	~~20~~
~~21~~	~~22~~	23	~~24~~	~~25~~	~~26~~	~~27~~	~~28~~	29	~~30~~
31	~~32~~	~~33~~	~~34~~	~~35~~	~~36~~	37	~~38~~	~~39~~	~~40~~
41	~~42~~	43	~~44~~	~~45~~	~~46~~	47	~~48~~	~~49~~	~~50~~
~~51~~	~~52~~	53	~~54~~	~~55~~	~~56~~	~~57~~	~~58~~	59	~~60~~
61	~~62~~	~~63~~	~~64~~	~~65~~	~~66~~	67	~~68~~	~~69~~	~~70~~
71	~~72~~	73	~~74~~	~~75~~	~~76~~	~~77~~	~~78~~	79	~~80~~
~~81~~	~~82~~	83	~~84~~	~~85~~	~~86~~	~~87~~	~~88~~	89	~~90~~
~~91~~	~~92~~	~~93~~	~~94~~	~~95~~	~~96~~	97	~~98~~	~~99~~	~~100~~
101									

Thus, all the multiples of $2, 3, 5, 7 < \sqrt{101}$ are sifted away. The number 53 is prime because it remained. Thus the primes up to 101 are 2, 3, 5, 7, 11, 13, 17, 19, 23, 29, 31, 37, 41, 43, 47, 53, 59, 61, 67, 71, 73, 79, 83, 89, 97, 101.

This procedure is the basis of sieve theory, which has been developed to provide estimates for the number of primes satisfying given conditions.

II. Some Fundamental Theorems on Congruences

In the next section, I intend to describe some classical methods to test primality and to find factors. They rely on theorems on congruences, especially Fermat's little theorem, the old theorem of Wilson, as well as Euler's generalization of Fermat's theorem. I shall also include a subsection on quadratic residues, a topic of central importance, which is also related with primality testing, as I shall have occasion to indicate.

A. FERMAT'S LITTLE THEOREM AND PRIMITIVE ROOTS MODULO A PRIME

Fermat's Little Theorem. *If p is a prime number and if a is an integer, then $a^p \equiv a \pmod{p}$. In particular, if p does not divide a then $a^{p-1} \equiv 1 \pmod{p}$.*

Euler published the first proof of Fermat's little theorem.

Proof. It is true for $a = 1$. Assuming that it is true for a, then, by induction, $(a + 1)^p \equiv a^p + 1 \equiv a + 1 \pmod{p}$. So the theorem is true for every natural number a.   □

The above proof required only the fact that if p is a prime number and if $1 \le k \le p - 1$, then the binomial coefficient $\binom{p}{k}$ is a multiple of p.

Note the following immediate consequence: if $p \nmid a$ and p^n is the highest power of p dividing $a^{p-1} - 1$, then p^{n+e} is the highest power of p dividing $a^{p^e(p-1)} - 1$ (where $e \ge 1$); in this statement, if $p = 2$, then n must be at least 2.

It follows from the theorem that for any integer a, which is not a multiple of the prime p, there exists the smallest exponent $h \ge 1$, such that $a^h \equiv 1 \pmod{p}$. Moreover, $a^k \equiv 1 \pmod{p}$ if and only if h divides k; in particular, h divides $p - 1$. This exponent h is called the *order of a modulo p*. Note that $a \bmod p$, $a^2 \bmod p$, ..., $a^{h-1} \bmod p$, and $1 \bmod p$ are all distinct.

It is a basic fact that for every prime p there exists at least one integer g, not a multiple of p, such that the order of g modulo p is equal to $p-1$. Then, the set $\{\, 1 \bmod p, g \bmod p, g^2 \bmod p, \ldots, g^{p-2} \bmod p \,\}$ is equal to the set $\{\, 1 \bmod p, 2 \bmod p, \ldots, (p - 1) \bmod p \,\}$.

Every integer g, $1 \le g \le p - 1$, such that $g \bmod p$ has order $p - 1$, is called a *primitive root modulo p*. I note this proposition:

Let p be any odd prime, $k \ge 1$, and $S = \sum_{j=1}^{p-1} j^k$. Then

$$S \equiv \begin{cases} -1 \bmod p, & \text{when } p - 1 | k, \\ 0 \bmod p, & \text{when } p - 1 \nmid k. \end{cases}$$

Proof. Indeed, if $p - 1$ divides k, then $j^k \equiv 1 \pmod{p}$ for $j = 1, 2, \ldots, p - 1$; so $S \equiv p - 1 \equiv -1 \pmod{p}$. If $p - 1$ does not divide k, let g be a primitive root modulo p. Then $g^k \not\equiv 1 \pmod{p}$. Since the sets of residue classes $\{\, 1 \bmod p, 2 \bmod p, \ldots, (p-1) \bmod p \,\}$ and $\{\, g \bmod p, 2g \bmod p, \ldots, (p-1)g \bmod p \,\}$ are the same, then

$$g^k S \equiv \sum_{j=1}^{p-1} (gj)^k \equiv \sum_{j=1}^{p-1} j^k \equiv S \pmod{p}.$$

Hence $(g^k - 1)S \equiv 0 \pmod{p}$ and, since p does not divide $g^k - 1$, then $S \equiv 0 \pmod{p}$. $\square$

The determination of a primitive root modulo p may be effected by a simple method indicated by Gauss in articles 73, 74 of *Disquisitiones Arithmeticae*.

Proceed as follows:

Step 1. Choose any integer a, $1 < a < p$, for example, $a = 2$, and write the residues modulo p of a, a^2, a^3, ... Let t be the smallest exponent such that $a^t \equiv 1 \pmod{p}$. If $t = p-1$, then a is a primitive root modulo p. Otherwise, proceed to the next step.

Step 2. Choose any number b, $1 < b < p$, such that $b \not\equiv a^i \pmod{p}$ for $i = 1, \ldots, t$; let u be the smallest exponent such that $b^u \equiv 1 \pmod{p}$. It is simple to see that u cannot be a factor of t, otherwise $b^t \equiv 1 \pmod{p}$; but $1, a, a^2, \ldots, a^{t-1}$ are t pairwise incongruent solutions of the congruence $X^t \equiv 1 \pmod{p}$; so they are all the possible solutions, and therefore $b \equiv a^m \pmod{p}$, for some m, $0 \leq m \leq t - 1$, which is contrary to the hypothesis. If $u = p - 1$, then b is a primitive root modulo p. If $u \neq p - 1$, let v be the least common multiple of t, u; so $v = mn$ with m dividing t, n dividing u, and $\gcd(m, n) = 1$. Let $a' \equiv a^{t/m} \pmod{p}$, $b' \equiv b^{u/n} \pmod{p}$ so $c = a'b'$ has order $mn = v$ modulo p. If $v = p - 1$, then c is a primitive root modulo p. Otherwise, proceed to the next step, which is similar to step 2.

Note that $v > t$, so in each step either one reaches a primitive root modulo p, or one constructs an integer with a bigger order modulo p.

The process must stop; one eventually reaches an integer with order $p - 1$ modulo p, that is, a primitive root modulo p.

Gauss also illustrated the procedure with the example $p = 73$, and found that $g = 5$ is a primitive root modulo 73.

The above construction leads to a primitive root modulo p, but not necessarily to the smallest integer g_p, $1 < g_p < p$, which is a primitive root modulo p.

The determination of g_p is done by trying successively the various integers $a = 2, 3, \ldots$ and computing their orders modulo p. There is no uniform way of predicting, for all primes p, which is the smallest primitive root modulo p. However, several results were known about the size of g_p. In 1944, Pillai proved that there exist infinitely many primes p, such that $g_p > C \log \log p$ (where C is a positive constant). In particular $\limsup_{p \to \infty} g_p = \infty$. A few years later, using a very deep theorem of Linnik (see Chapter 4) on primes in arithmetic progressions, Fridlender (1949), and independently Salié (1950), proved that $g_p > C \log p$ (for some constant C) and infinitely many primes p. On the other hand, g_p does not grow too fast, as proved by Burgess in 1962:

$$g_p \le C p^{1/4 + \varepsilon}$$

(for $\varepsilon > 0$, a constant $C > 0$, and p sufficiently large).

Grosswald made Burgess' result explicit in 1981: if $p > e^{e^{24}}$ then $g_p < p^{0.499}$.

The proof of the weaker result (with $\frac{1}{2}$ in place of $\frac{1}{4}$), attributed to Vinogradov, is in Landau's *Vorlesungen über Zahlentheorie*, Part VII, Chapter 14 (see General References).

The proof of the following result is elementary (problem proposed by Powell in 1983, solution by Kearnes in 1984):

For any positive integer M, there exist infinitely many primes p such that $M < g_p < p - M$.

A simple glance at the table suggests the following problem.

Is 2 a primitive root for infinitely many primes? More generally, if the integer $a \ne \pm 1$ is not a square, is it a primitive root modulo infinitely many primes?

This is a difficult problem and I shall return to it in Chapter 4.

Table

p	g_p	p	g_r
2	1	89	3
3	2	97	5
5	2	101	2
7	3	103	5
11	2	107	2
13	2	109	6
17	3	113	3
19	2	127	3
23	5	131	2
29	2	137	3
31	3	139	2
37	2	149	2
41	6	151	6
43	3	157	5
47	5	163	2
53	2	167	5
59	2	173	2
61	2	179	2
67	2	181	2
71	7	191	19
73	5	193	5
79	3	197	2
83	2	199	3

B. The Theorem of Wilson

Wilson's Theorem. *If p is a prime number, then*

$$(p-1)! \equiv -1 \pmod{p}.$$

Proof. This is just a corollary of Fermat's little theorem. Indeed, $1, 2, \ldots, p-1$ are roots of the congruence $X^{p-1} - 1 \equiv 0 \pmod{p}$. But a congruence modulo p cannot have more roots than its degree. Hence,

$$X^{p-1} - 1 \equiv (X-1)(X-2) \cdots (X-(p-1)) \pmod{p}.$$

Comparing the constant terms, $-1 \equiv (-1)^{p-1}(p-1)! = (p-1)!$ (mod p). (This is also true if $p = 2$.)   □

Wilson's theorem gives a characterization of prime numbers. Indeed, if $N > 1$ is a natural number that is not a prime, then $N = mn$, with $1 < m, n < N - 1$, so m divides N and $(N - 1)!$, and therefore $(N - 1)! \not\equiv -1 \pmod{N}$.

However, Wilson's characterization of the prime numbers is not of practical value to test the primality of N, since there is no known algorithm to rapidly compute $N!$, say, in $\log N$ steps.

C. THE PROPERTIES OF GIUGA AND OF WOLSTENHOLME

Now, I shall consider other properties that are satisfied by prime numbers.

First, I note that if p is a prime, then by Fermat's little theorem (as already indicated)

$$1^{p-1} + 2^{p-1} + \cdots + (p - 1)^{p-1} \equiv -1 \pmod{p}.$$

In 1950, Giuga asked whether the converse is true: If $n > 1$ and n divides $1^{n-1} + 2^{n-1} + \cdots + (n - 1)^{n-1} + 1$, then is n a prime number?

Giuga verified that if $n < 10^{1000}$ and satisfies the above congruence, then it is a prime. Bedocchi (1985) extended this verification for all $n < 10^{1700}$. In their computations, they made use of the following characterization:

n satisfies Giuga's condition if and only if, for every prime p dividing n, $p^2(p - 1)$ divides $n - p$.

From this, it follows that if p divides n, then $p - 1$ divides $n - 1$. It will be seen in Section IX that this latter condition implies that $a^n \equiv a \pmod{n}$ for every integer $a \geq 1$, so either n is a prime or it is a Carmichael number. A final attack on the problem is needed. It may be not as simple as one would like. I did not try, but I know of two good mathematicians who tried, but failed. Don't ask me their names. It is all to their credit to hope to solve a problem like this, which has, at first sight, no reason to have a positive or negative answer.

Incidentally, this is an instance where, instead of more calculations, it would be preferable to let the brain work.

Now, I shall discuss another property of prime numbers.

In 1862, Wolstenholme proved the following interesting result: If p is a prime, $p \geq 5$, then the numerator of

$$1 + \frac{1}{2} + \frac{1}{3} + \cdots + \frac{1}{p - 1}$$

is divisible by p^2, and the numerator of

$$1 + \frac{1}{2^2} + \frac{1}{3^2} + \cdots + \frac{1}{(p-1)^2}$$

is divisible by p.

For a proof, see Hardy & Wright (1938, p. 88, General References). Based on this property, it is not difficult to deduce that if $n \geq 5$ is a prime number, then

$$\binom{2n-1}{n-1} \equiv 1 \pmod{n^3}.$$

Is the converse true? This question, still unanswered today, has been asked by Jones for a few years. An affirmative reply would provide an interesting and formally simple characterization of prime numbers.

D. THE POWER OF A PRIME DIVIDING A FACTORIAL

In 1808, Legendre determined the exact power p^m of the prime p that divides a factorial $a!$ (so p^{m+1} does not divide $a!$).

There is a very nice expression of m in terms of the p-adic development of a:

$$a_k = ap^k + a_{k-1}p^{k-1} + \cdots + a_1 p + a_0$$

where $p^k \leq a < p^{k+1}$ and $0 \leq a_i \leq p-1$ (for $i = 0, 1, \ldots, k$). The integers $a_0, a_1, \ldots, a_k$ are the digits of a in base p.

For example, in base 5, $328 = 2 \times 5^3 + 3 \times 5^2 + 3$, so the digits of 328 in base 5 are $2, 3, 0, 3$. Using the above notation,

$$m = \sum_{i=0}^{\infty} \left[\frac{a}{p^i} \right] = \frac{a - (a_0 + a_1 + \cdots + a_k)}{p-1}.$$

Proof. By definition $a! = p^m b$, where $p \nmid b$.

Let $a = q_1 p + r_1$ with $0 \leq q_1$, $0 \leq r_1 < p$; so $q_1 = [a/p]$. The multiples of p, not bigger than a are $p, 2p, \ldots, q_1 p \leq a$. So $p^{q_1}(q_1!) = p^m b'$, where $p \nmid b'$. Thus $q_1 + m_1 = m$, where p^{m_1} is the exact power of p which divides $q_1!$. Since $q_1 < a$, by induction,

$$m_1 = \left[\frac{q_1}{p} \right] + \left[\frac{q_1}{p^2} \right] + \left[\frac{q_1}{p^2} \right] + \cdots.$$

But

$$\left[\frac{q_1}{p^i}\right] = \left[\frac{[a/p]}{p^i}\right] = \left[\frac{a}{p^{i+1}}\right],$$

as may be easily verified. So

$$m = \left[\frac{a}{p}\right] + \left[\frac{a}{p^2}\right] + \left[\frac{a}{p^3}\right] + \cdots.$$

Now, I derive the second expression, involving the p-adic digits of $a = a_k p^k + \cdots + a_1 p + a_0$. Then

$$\left[\frac{a}{p}\right] = a_k p^{k-1} + \cdots + a_1,$$

$$\left[\frac{a}{p^2}\right] = a_k p^{k-2} + \cdots + a_2,$$

$$\vdots$$

$$\left[\frac{a}{p^k}\right] = a_k.$$

So

$$\sum_{i=0}^{\infty} \left[\frac{a}{p^i}\right] = a_1 + a_2(p+1) + a_3(p^2 + p + 1) + \cdots +$$

$$a_k(p^{k-1} + p^{k-2} + \cdots + p + 1)$$

$$= \frac{1}{p-1}\{ a_1(p-1) + a_2(p^2 - 1) + \cdots + a_k(p^k - 1) \}$$

$$= \frac{1}{p-1}\{ a - (a_0 + a_1 + \cdots + a_k) \}.$$

$\square$

In 1852, Kummer used Legendre's result to determine the exact power p^m of p dividing a binomial coefficient

$$\binom{a+b}{a} = \frac{(a+b)!}{a! b!},$$

where $a \geq 1$, $b \geq 1$ This discovery is also credited to Lucas, who obtained the result independently, but later (before 1871).

Let

$$a = a_0 + a_1 p + \cdots + a_t p^t,$$

$$b = b_0 + b_1 p + \cdots + b_t p^t,$$

where $0 \leq a_i \leq p-1$, $0 \leq b_i \leq p-1$, and either $a_t \neq 0$ or $b_t \neq 0$. Let $S_a = \sum_{i=0}^{t} a_i$, $S_b = \sum_{i=0}^{t} b_i$ be the sums of p-adic digits of a, b. Let c_i, $0 \leq c_i \leq p-1$, and $\varepsilon_i = 0$ or 1, be defined successively as follows:

$$
\begin{aligned}
a_0 + b_0 &= \varepsilon_0 p + c_0, \\
\varepsilon_0 + a_1 + b_1 &= \varepsilon_1 p + c_1, \\
\varepsilon_1 + a_2 + b_2 &= \varepsilon_2 p + c_2, \\
&\vdots \\
\varepsilon_{t-1} + a_t + b_t &= \varepsilon_t p + c_t.
\end{aligned}
$$

Multiplying these equations successively by 1, p, p^2, ... and adding them:

$$
\begin{aligned}
a + b + \varepsilon_0 p + \varepsilon_1 p^2 + \cdots \quad + \quad \varepsilon_{t-1} p^t &= \varepsilon_0 p + \varepsilon_1 p^2 + \cdots \\
&+ \quad \varepsilon_{t-1} p^t + \varepsilon_t p^{t+1} + c_0 + c_1 p + \cdots + c_t p^t.
\end{aligned}
$$

So, $a + b = c_0 + c_1 p + \cdots + c_t p^t + \varepsilon_t p^{t+1}$, and this is the expression of $a + b$ in the base p. Similarly, by adding those equations:

$$
S_a + S_b + (\varepsilon_0 + \varepsilon_1 + \cdots + \varepsilon_{t-1}) = (\varepsilon_0 + \varepsilon_1 + \cdots + \varepsilon_t)p + S_{a+b} - \varepsilon_t.
$$

By Legendre's result

$$
\begin{aligned}
(p-1)m &= (a+b) - S_{a+b} - a + S_a - b + S_b \\
&= (p-1)(\varepsilon_0 + \varepsilon_1 + \cdots + \varepsilon_t).
\end{aligned}
$$

Hence, the result of Kummer:

The exponent of the exact power of p dividing $\binom{a+b}{a}$ is equal to $\varepsilon_0 + \varepsilon_1 + \cdots + \varepsilon_t$, which is the number of "carry-overs" when performing the addition of a, b, written in the base p.

The results of Legendre and Kummer have found many applications, in p-adic analysis, and also, for example, in Chapter 3, Section III.

E. THE CHINESE REMAINDER THEOREM

Even though my paramount interest is on prime numbers, there is no way to escape also dealing with arbitrary integers—which essentially amounts, in many questions, to the simultaneous consideration of several primes, because of the decomposition of integers into the product of prime powers.

One of the keys connecting results for integers n and for their prime power factors is very old; indeed, it was known to the ancient Chinese and, therefore, it is called the Chinese remainder theorem. I'm sure that every one of my readers knows it already:

If $n_1, n_2, \ldots, n_k$ are pairwise relatively prime positive integers, and if $a_1, a_2, \ldots, a_k$ are any integers, then there exists an integer a such that

$$
\begin{cases}
a \equiv a_1 \pmod{n_1} \\
a \equiv a_2 \pmod{n_2} \\
\vdots \\
a \equiv a_k \pmod{n_k}.
\end{cases}
$$

Another integer a' satisfies also the same congruences as a if and only if $a \equiv a' \pmod{n_1 n_2 \cdots n_k}$. So, there exists a unique integer a, as above, with $0 \le a < n_1 n_2 \cdots n_k$.

The proof is indeed very simple; it is in many books and also in a short note by Mozzochi (1967).

There are many uses for the Chinese remainder theorem. It was in fact in this way that Chinese generals used to count the number of their soldiers:

Lineup 7 by 7!   (Not factorial of 7, but a SCREAMED
                   military command.)
Lineup 11 by 11!
Lineup 13 by 13!
Lineup 17 by 17!

and they could go in this way, just counting the remainders.

In his historical research, A. Zachariou has uncovered evidence that the Greeks, even before the Chinese, knew the Chinese remainder theorem. I wonder if it was also used for the purpose of counting soldiers...

Here is another application: if $n = p_1 p_2 \cdots p_t$ is a product of distinct primes, if g_i is a primitive root modulo p_i, if g is such that $1 \le g \le n - 1$ and $g \equiv g_i \pmod{p_i}$ for every $i = 1, \ldots, t$, then g is a common primitive root modulo every p_i.

F. EULER'S FUNCTION

Euler generalized Fermat's little theorem by introducing the *totient* or *Euler's function*.

For every $n \geq 1$, let $\phi(n)$ denote the number of integers a, $1 \leq a < n$, such that $\gcd(a, n) = 1$. Thus, if $n = p$ is a prime, then $\phi(p) = p - 1$; also

$$\phi(p^k) = p^{k-1}(p - 1) = p^k \left(1 - \frac{1}{p} \right).$$

Moreover, if $m, n \geq 1$ and $\gcd(m, n) = 1$, then $\phi(mn) = \phi(m)\phi(n)$, that is, ϕ is a multiplicative function. Hence, for any integer $n = \prod_p p^k$ (product for all primes p dividing n, and $k \geq 1$), then

$$\phi(n) = \prod_p p^{k-1}(p - 1) = n \prod_p \left(1 - \frac{1}{p} \right).$$

Another simple property is: $n = \Sigma_{d|n}\phi(d)$.

Euler proved the following:

If $\gcd(a, n) = 1$, then $a^{\phi(n)} \equiv 1 \pmod{n}$.

Proof. Let $r = \phi(n)$ and let $b_1, \ldots, b_r$ be integers, pairwise incongruent modulo n, such that $\gcd(b_i, n) = 1$ for $i = 1, \ldots, r$.

Then $ab_1, \ldots, ab_r$ are again pairwise incongruent modulo n and $\gcd(ab_i, n) = 1$ for $i = 1, \ldots, r$. Therefore, the sets $\{\, b_1 \bmod n, \ldots, b_r \bmod n \,\}$ and $\{\, ab_1 \bmod n, \ldots, ab_r \bmod n \,\}$ are equal. Now,

$$a^r \prod_{i=1}^{r} b_i \equiv \prod_{i=1}^{r} ab_i \equiv \prod_{i=1}^{r} b_i \pmod{n}.$$

Hence,

$$(a^r - 1) \prod_{i=1}^{r} b_i \equiv 0 \pmod{n} \quad \text{and so} \quad a^r \equiv 1 \pmod{n}.$$

$\square$

Just like for Fermat's little theorem, it follows also from Euler's theorem that there exists the smallest positive exponent e such that $a^e \equiv 1 \pmod{n}$. It is called the *order of a modulo n*. If n is a prime number, this definition coincides with the previous one. Note also that $a^m \equiv 1$ (mod n) if and only if m is a multiple of the order e of a mod n; thus, in particular, e divides $\phi(n)$.

Once again, it is natural to ask: Given $n > 2$ does there always exist an integer a, relatively prime to n, such that the order of a mod n is equal to $\phi(n)$? Recall that when $n = p$ is a prime, such numbers exist, namely, the primitive roots modulo p. If $n = p^e$, a power of an odd prime, it is also true. More precisely, the following assertions are equivalent:

(i) g is a primitive root modulo p and $g^{p-1} \not\equiv 1$ (mod p^2);

(ii) g is a primitive root modulo p^2;

(iii) for every $e \geq 2$, g is a primitive root modulo p^e.

Note that 10 is a primitive root modulo 487, but $10^{486} \equiv 1$ (mod 487^2), so 10 is not a primitive root modulo 487^2. This is the smallest example illustrating this possibility, when the base is 10. Another example is 14 modulo 29.

However, if n is divisible by $4p$, or pq, where p, q are distinct odd primes, then there is no number a, relatively prime to n, with order equal to $\phi(n)$. Indeed, it is easy to see that the order of a mod n is at most equal to $\lambda(n)$, where $\lambda(n)$ is the following function, defined by Carmichael in 1912:

$\lambda(1) = 1$, $\lambda(2) = 1$, $\lambda(4) = 2$, $\lambda(2^r) = 2^{r-2}$ (for $r \geq 3$),
$\lambda(p^r) = p^{r-1}(p-1) = \phi(p^r)$ for any odd prime p and $r \geq 1$,

$$\lambda(2^r p_1^{r_1} p_2^{r_2} \cdots p_s^{r_s}) = \ell\text{cm}\{\lambda(2^r), \lambda(p_1^{r_1}), \ldots, \lambda(p_s^{r_s})\}$$

(ℓcm denotes the least common multiple).

Note that $\lambda(n)$ divides $\phi(n)$, but may be smaller, and that there is an integer a, relatively prime to n, with order of a mod n equal to $\lambda(n)$.

I shall use this opportunity to study Euler's function in more detail. First I shall consider Lehmer's problem, and thereafter the values of ϕ, the valence, the values avoided, the average of the function, etc.

Lehmer's problem.

Recall that if p is a prime, then $\phi(p) = p - 1$. In 1932, Lehmer asked whether there exists any composite integer n such that $\phi(n)$ divides $n - 1$. This question remains open.

What can one say, anyway, when it is not possible to solve the problem? Only that the existence of composite integers n, for which $\phi(n)$ divides $n - 1$, is unlikely, for various reasons:

(a) any such number must be very large (if it exists at all);

(b) any such number must have many prime factors (if it exists at all);

(c) the number of such composite numbers, smaller than any given real number x, is bounded by a very small function of x.

Thus, Lehmer showed in 1932 that if n is composite and $\phi(n)$ divides $n - 1$, then n is odd and square-free, and the number of its distinct prime factors is $w(n) \geq 7$. Subsequent work by Schuh (1944) gave $w(n) \geq 11$.

Record

To date, the best results are $n > 10^{20}$, $w(n) \geq 14$ by Cohen & Hagis (1980) and if $30 \nmid n$, then $w(n) \geq 26$ by Wall (1980), while if $3 | n$, then Lieuwens' result (1970) is still the best: $w(n) \geq 213$ and $n \geq 5.5 \times 10^{570}$.

Moreover, in 1977 Pomerance showed that for every sufficiently large positive real number x, the number $L(x)$ of composite n such that $\phi(n) | n - 1$ and $n \leq x$, satisfies

$$L(x) \leq x^{1/2} (\log x)^{3/4}.$$

And, if $w(n) = k$, then $n < k^{2^k}$.

Values of Euler's function.

Not every even integer $n \geq 1$ is a value of Euler's function—a fact which is not difficult to establish. For example, Schinzel showed in 1956 that, for every $k \geq 1$, 2×7^k is not a value of Euler's function. In 1976, Mendelsohn showed that there exist infinitely many primes p such that, for every $k \geq 1$, $2^k p$ is not a value of the function ϕ.

Moreover, for every $k \geq 1$ there exists n such that $\phi(n) = k!$; this was proved by Gupta in 1950.

The next results tell how erratic is the behaviour of Euler's function.

Sierpiński and Schinzel (see Schinzel, 1954) showed, improving a previous result of Somayajulu (1950):

The set of all numbers $\phi(n+1)/\phi(n)$ is dense in the set of all real positive numbers.

Schinzel & Sierpiński (1954) and Schinzel (1954) also proved the following: For every m, $k \geq 1$, there exist n, $h \geq 1$ such that

$$\frac{\phi(n+i)}{\phi(n+i-1)} > m \quad \text{and} \quad \frac{\phi(h+i-1)}{\phi(h+i)} > m$$

for $i = 1, 2, \ldots, k$.

It is also true that the set of all numbers $\phi(n)/n$ is dense in the interval $(0, 1)$.

The valence of Euler's function.

Now I shall examine the "valence" of Euler's function; in other words, how often a value $\phi(n)$ is assumed. In order to explain the results in a systematic way, it is better to introduce some notation. If $m \geq 1$, let

$$N_\phi(m) = \#\{\, n \geq 1 \mid \phi(n) = m \,\}.$$

What are the possible values of $N_\phi(m)$? I have already said that there are infinitely many even integers m for which $N_\phi(m) = 0$. It is also true that if $m = 2 \times 3^{6k+1}$ ($k \geq 1$), then $\phi(n) = m$ exactly when $n = 3^{6k+2}$ or $n = 2 \times 3^{6k+2}$. Hence, there are infinitely many integers m such that $N_\phi(m) = 2$.

It is not difficult to show that $N_\phi(m) \neq \infty$ for every $m \geq 1$.

Schinzel gave a simpler proof (in 1956) of the following result of Pillai (1929):

$$\sup\{\, N_\phi(m) \,\} = \infty.$$

In other words, for every $k \geq 1$ there exists an integer m_k such that there exists at least k integers n with $\phi(n) = m_k$.

Carmichael's conjecture.

The conjecture that dominates the study of the valence of ϕ was proposed by Carmichael in 1922: N_ϕ does not assume the value 1.

In other words, given $n \geq 1$, there exists $n' \geq 1$, $n' \neq n$, such that $\phi(n') = \phi(n)$.

A concise update on Carmichael's conjecture, written by Wagon, has appeared in *The Mathematical Intelligencer*, 1986.

This conjecture was studied by Klee, who showed in 1947 that it holds for every integer n such that $\phi(n) < 10^{400}$. Klee's method was improved by Masai & Valette (1982), and now the statement is known to be true for all n such that $\phi(n) < 10^{10000}$.

Pomerance has shown (1974) the following: Suppose that m is a natural number such that if p is any prime and $p - 1$ divides $\phi(m)$, then p^2 divides m. Then $N_\phi(\phi(m)) = 1$.

Of course, if there exists a number m satisfying the above condition, then Carmichael's conjecture would be false. However, the existence of such a number m is far from established, and perhaps unlikely.

Finally, in variance with Carmichael's conjecture, it is reasonable to expect that every $s \neq 1$ is a value of N_ϕ; this was conjectured by Sierpiński. As a matter of fact, I shall indicate in Chapter 6 that this statement follows from an unproved and very interesting hypothesis.

The growth of $\phi(n)$.

I haven't yet considered the growth of the function ϕ. Since $\phi(p) = p - 1$ for every prime p, then $\limsup \phi(n) = \infty$. Similarly, from $\phi(p) = p - 1$, $\limsup \phi(n)/n = 1$.

I shall postpone the indication of other results about the growth of ϕ until Chapter 4: they depend on methods that will be discussed in that chapter.

G. SEQUENCES OF BINOMIALS

The preceding considerations referred to congruences modulo a given integer $n > 1$, and a was any positive integer relatively prime to n.

Another point of view is very illuminating. This time, let $a > 1$ be given, and consider the sequence of integers $a^n - 1$ (for $n \geq 1$), as well as the companion sequence of integers $a^n + 1$ (for $n \geq 1$). More generally, if $a > b \geq 1$ with $\gcd(a, b) = 1$, one may consider the sequences $a^n - b^n$ $(n \geq 1)$ and $a^n + b^n$ $(n \geq 1)$.

A first natural question, with an immediate answer, is the following: to determine all primes p, such that there exists $n \geq 1$ for which

p divides $a^n - b^n$ (respectively, $a^n + b^n$). These are primes not dividing ab because a, b are relatively prime. Conversely, if $p \nmid ab$, if $bb' \equiv 1$ (mod p) and n is the order of ab' mod p (resp. of $-ab'$ mod p), then p divides $a^n - b^n$ (resp. $a^n + b^n$).

If $n \geq 1$ is the smallest integer such that p divides $a^n - b^n$ (resp. $a^n + b^n$), then p is called a *primitive prime factor of* $a^n - b^n$ (resp. $a^n + b^n$). In this case, by Fermat's little theorem, n divides $p - 1$; this was explicitly observed by Legendre.

So, every prime $p \nmid ab$ appears as a primitive factor of some binomial. Does, conversely, every binomial have a primitive factor?

In 1892, Zsigmondy proved the following theorem, which is very interesting and has many applications:

If $a > b \geq 1$ and $\gcd(a, b) = 1$, then every number $a^n - b^n$ has a primitive prime factor—the only exceptions being $a = 2$, $b = 1$, $n = 6$; $a + b$ power of 2 (hence a, b are odd), $n = 2$.

Equally, if $a > b \geq 1$, then every number $a^n + b^n$ has a primitive prime factor—with the exception of $2^3 + 1 = 9$.

The special case, where $b = 1$, had been proved by Bang in 1886.

Later, this theorem, or Bang's special case was proved again, sometimes unknowingly, by a long list of mathematicians: Birkhoff & Vandiver (1904), Carmichael (1913), Kanold (1950), Artin (1955), Lüneburg (1981), and probably others.

The proof is definitely not obvious; however, it is very easy to write up such sequences and watch the successive appearance of new primitive prime factors.

It is interesting to consider the primitive part t_n^* of $a^n - b^n$; namely, write $a^n - b^n = t_n^* t_n'$ with $\gcd(t_n^*, t_n') = 1$ and a prime p divides t_n^* if and only if p is a primitive factor of $a^n - b^n$.

By experimenting numerically with sequences $a^n - b^n$, it is observed that, apart from a few initial terms, t_n^* is composite.

In fact, let $k(m)$ denote the square-free kernel of m, that is, m divided by its largest square factor. Let

$$e = \begin{cases} 1 & \text{if } k(ab) \equiv 1 (\mathrm{mod}\, 4) \\ 2 & \text{if } k(ab) \equiv 2 \text{ or } 3 (\mathrm{mod}\, 4). \end{cases}$$

In 1962, Schinzel showed that $\frac{n}{ek(ab)}$ is an odd integer, then $a^n - b^n$ has at least two distinct primitive prime factors, with the following exceptions, when $n > 1$:

$$(a,b) = (2,1) \quad : \quad n=4,\ 12,\ 20;$$
$$(a,b) = (3,1) \quad : \quad n=6;$$
$$(a,b) = (3,2) \quad : \quad n=12;$$
$$(a,b) = (4,1) \quad : \quad n=3;$$
$$(a,b) = (4,3) \quad : \quad n=6.$$

Schinzel also proved that if $ab = c^h$ with $h \geq 3$, then there are infinitely many n such that the primitive part of $a^n - b^n$ has at least three prime factors.

For the sequence of binomials $a^n + b^n$, it follows at once: if $\frac{n}{\ell k(ab)}$ is odd, and $n > 10$, then the primitive part of $a^n + b^n$ has at least two primitive prime factors. Just note that each primitive prime factor of $a^{2n} - b^{2n}$ is also a primitive prime factor of $a^n + b^n$.

Here are some questions that are very difficult to answer:

> Are there infinitely many n such that the primitive part of $a^n - b^n$ is prime?

> Are there infinitely many n such that the primitive part of $a^n - b^n$ is square-free?

And how about the seemingly easier questions:

> Are there infinitely many n such that the primitive part t_n^* of $a^n - b^n$ has a prime factor p such that p_n^2 does not divide $a^n - b^n$?

> Are there infinitely many n such that t_n^* has a square-free kernel $k(t_n^*) \neq 1$?

These questions, for the special case when $b = 1$, are ultimately related, in a very surprising way, to Fermat's last theorem!

H. QUADRATIC RESIDUES

In the study of quadratic diophantine equations, developed by Fermat, Euler, Legendre, Gauss, it was very important to determine when an integer a is a square modulo a prime $p > 2$.

If $p > 2$ does not divide a and if there exists an integer b such that $a \equiv b^2 \pmod{p}$, then a is called a *quadratic residue modulo p*; otherwise, it is a *nonquadratic residue modulo p*.

Legendre introduced the following practical notation:

$$\left(\frac{a}{p}\right) = (a/p) = \begin{cases} +1 & \text{if } a \text{ is a quadratic residue modulo } p, \\ -1 & \text{otherwise.} \end{cases}$$

It is also convenient to define $(a/p) = 0$ when p divides a.

I shall now indicate the most important properties of the Legendre symbol. References are plentiful—practically every book in elementary number theory.

If $a \equiv a' \pmod{p}$, then

$$\left(\frac{a}{p}\right) = \left(\frac{a'}{p}\right).$$

For any integers a, a':

$$\left(\frac{aa'}{p}\right) = \left(\frac{a}{p}\right)\left(\frac{a'}{p}\right).$$

So, for the computation of the Legendre symbol, it suffices to calculate (q/p), where $q = -1$, 2, or any odd prime different from p.

Euler proved the following congruence:

$$\left(\frac{a}{p}\right) \equiv a^{(p-1)/2} \pmod{p}.$$

In particular,

$$\left(\frac{-1}{p}\right) = \begin{cases} +1 & \text{when } p \equiv 1 \pmod{4}, \\ -1 & \text{when } p \equiv -1 \pmod{4}, \end{cases}$$

and

$$\left(\frac{1}{p}\right) = \begin{cases} +1 & \text{when } p \equiv \pm 1 \pmod{8}, \\ -1 & \text{when } p \equiv \pm 3 \pmod{8}. \end{cases}$$

The computation of the Legendre symbol (q/p), for any odd prime $q \neq p$, can be performed with an easy, explicit and fast algorithm (needing only Euclidean division), by using Gauss' reciprocity law:

$$\left(\frac{p}{q}\right) = \left(\frac{q}{p}\right)(-1)^{\frac{p-1}{2} \times \frac{q-1}{2}}.$$

The importance of Legendre's symbol was such that it prompted Jacobi to consider the following generalization, now called Jacobi symbol. Again, references are abundant, for example, Grosswald's book (1966, second edition 1984), or, why not?, my own book (1972).

Let a be a nonzero integer, let b be an odd integer, such that $\gcd(a, b) = 1$. The Jacobi symbol (a/b) is defined as an extension of Legendre's symbol, in the following manner. Let $|b| = \prod_{p|b} p^{e_p}$ (with $e_p \geq 1$). Then

$$\left(\frac{a}{b}\right) = \left(\frac{a}{-b}\right) = \prod_{p|b} \left(\frac{a}{p}\right)^{e_p}.$$

Therefore, (a/b) is equal to $+1$ or -1. Note that

$$\left(\frac{a}{1}\right) = \left(\frac{a}{-1}\right) = +1.$$

Here are some of the properties of the Jacobi symbol (under the assumptions of its definition):

$$\left(\frac{aa'}{b}\right) = \left(\frac{a}{b}\right)\left(\frac{a'}{b}\right),$$

$$\left(\frac{a}{bb'}\right) = \left(\frac{a}{b}\right)\left(\frac{a}{b'}\right)$$

$$\left(\frac{-1}{b}\right) = (-1)^{(b-1)/2} = \begin{cases} +1 & \text{if } b \equiv 1 \pmod 4, \\ -1 & \text{if } b \equiv -1 \pmod 4, \end{cases}$$

$$\left(\frac{2}{b}\right) = (-1)^{(b-1)/8} = \begin{cases} +l & \text{if } b \equiv \pm 1 \pmod 8, \\ -1 & \text{if } b \equiv \pm 3 \pmod 8. \end{cases}$$

For the calculation of the Jacobi symbol, the key result is the reciprocity law, which follows easily from Gauss' reciprocity law for the Legendre symbol:

If a, b are relatively prime odd integers and $b \geq 3$, then

$$\left(\frac{a}{b}\right) = (-1)^{\frac{a-1}{2} \times \frac{b-1}{2}} \left(\frac{b}{a}\right).$$

If $b \geq 3$ and a is a square modulo b, then

$$\left(\frac{a}{b}\right) = +1.$$

III. Classical Primality Tests Based on Congruences

After the discussion of the theorems of Fermat, Wilson, and Euler, I am ready. For me, the classical primality tests based on congruences are those indicated by Lehmer, extending or using previous tests by Lucas, Pocklington, and Proth. I reserve another section for tests based on recurring sequences.

Wilson's theorem, which characterizes prime numbers, might seem very promising. But, it has to be discarded as a practical test, since the computation of factorials is very time consuming.

Fermat's little theorem says that if p is a prime and a is any natural number not a multiple of p, then $a^{p-1} \equiv 1 \pmod{p}$. However, I note right away that a crude converse of this theorem is not true—because there exist composite integers N, and $a \geq 2$, such that $a^{N-1} \equiv 1 \pmod{N}$. I shall devote Section VIII to the study of these numbers, which are very important in primality questions.

Nevertheless, a true converse of Fermat's little theorem was discovered by Lucas in 1876. It says:

Test 1. Let $N > 1$. Assume that there exists an integer $a > 1$ such that:

(i) $a^{N-1} \equiv 1 \pmod{N}$,

(ii) $a^m \not\equiv 1 \pmod{N}$, for $m = 1, 2, \ldots, N - 2$.

Then N is a prime.

Defect of this test: it might seem perfect, but it requires $N - 2$ successive multiplications by a, and finding residues modulo N-too many operations.

Proof. It suffices to show that every integer m, $1 \leq m < N$, is prime to N, that is, $\phi(N) = N - 1$. For this purpose, it suffices to show that there exists a, $1 \leq a < N$, $\gcd(a, N) = 1$, such that the order of $a \bmod N$ is $N - 1$. This is exactly spelled out in the hypothesis.   □

In 1891, Lucas gave the following test:

Test 2. Let $N > 1$. Assume that there exists an integer $a > 1$ such that:

(i) $a^{N-1} \equiv 1 \pmod{N}$,

(ii) $a^m \not\equiv 1 \pmod{N}$ for every $m < N$, such that m divides $N-1$.

Then N is a prime.

Defect of this test: it requires the knowledge of all factors of $N-1$, thus it is only easily applicable when $N - 1$ can be factored, like $N = 2^n + 1$, or $N = 3 \times 2^n + 1$.

The proof of Test 2 is, of course, the same as that of Test 1.

In 1967, Brillhart & Selfridge made Lucas' test more flexible; see also the paper by Brillhart, Lehmer & Selfridge, in 1975:

Test 3. Let $N > 1$. Assume that for every prime factor q of $N - 1$ there exists an integer $a = a(q) > 1$ such that

(i) $a^{N-1} \equiv 1 \pmod{N}$,

(ii) $a^{(N-1)/q} \not\equiv 1 \pmod{N}$.

Then N is a prime.

Defect of this test: once again, it is necessary to know the prime factors of $N - 1$, but fewer congruences have to be satisfied.

An observant reader should note that, after all, to verify that $a^{N-1} \equiv 1 \pmod{N}$ it is necessary in particular to calculate, as one goes, the residue of a^n modulo N (for every $n \leq N - 1$), and so the first Lucas' criterion could have been used. The point is that there is a fast algorithm to find the power a^n, hence also $a^n \bmod N$, without computing all the preceding powers. It runs as follows.

Write the exponent n in base 2:

$$n = n_0 2^k + n_1 2^{k-1} + \cdots + n_{k-1} 2 + n_k,$$

where each n_i is equal to 0 or 1, and $n_0 = 1$.

Put $s_0 = n_0 = 1$ and if s_j has been calculated, let $s_{j+1} = 2s_j + n_{j+1}$. Let $r_j = a^{s_j}$. Then $r_{j+1} = r_j^2 a^{j+1}$; so $r_{j+1} = r_j^2$ when $n_{j+1} = 0$, or $r_{j+1} = ar_j^2$ when $n_{j+1} = 1$. Noting that $r_k = a^n$, then it is only necessary to perform $2k$ operations, which are either a squaring or a multiplication by a. If the computation is of $a^n \bmod N$, then it is even easier; at each stage r_j is to be replaced by its residue modulo N. Now, k is equal to

$$\left\lfloor \frac{\log n}{\log 2} \right\rfloor .$$

So, if $n = N - 1$, then only about

$$2 \left\lceil \frac{\log N}{\log 2} \right\rceil$$

operations are needed to find $a^{N-1} \bmod N$, and there is no need to compute all powers $a^n \bmod N$.

Why don't you try calculating $2^{1092} \bmod 1093^2$ in this way? You should find $2 \equiv 1 \pmod{1093^2}$—if you really succeed! This has nothing to do directly with primality—but it will appear much later, in Chapter 5.

I return to Brillhart & Selfridge's Test 3 and give its proof.

Proof of Test 3. It is enough to show that $\phi(N) = N - 1$, and since $\phi(N) \leq N - 1$, it suffices to show that $N - 1$ divides $\phi(N)$. If this is false, there exists a prime q and $r \geq 1$ such that q^r divides $N - 1$, but q^r does not divide $\phi(N)$. Let $a = a(q)$ and let e be the order of $a \bmod N$. Thus e divides $N - 1$ and e does not divide $(N - 1)/q$, so q^r divides e. Since $a^{\phi(N)} \equiv 1 \pmod{N}$, then e divides $\phi(N)$, so $q^r | \phi(N)$, which is a contradiction, and concludes the proof. □

In the section on Fermat numbers, I will derive Pepin's primality test for Fermat numbers, as a consequence of Test 3.

To make the primality tests more efficient, it is desirable to avoid the need to find all prime factors of $N - 1$. So there are tests that only require a partial factorization of $N - 1$. The basic result was proved by Pocklington in 1914, and it is indeed very simple:

Let $N - 1 = q^n R$, where q is a prime, $n \geq 1$, and q does not divide R. Assume that there exists an integer $a > 1$ such that:

(i) $a^{N-1} \equiv 1 \pmod{N}$,

(ii) $\gcd(a^{(N-1)/q} - 1, N) = 1$.

Then each prime factor of N is of the form $mq^n + 1$, with $m \geq 1$.

Proof. Let p be a prime factor of N, and let e be the order of $a \bmod p$, so e divides $p - 1$; by condition (ii), e cannot divide $(N - 1)/q$, because p divides N; hence, q does not divide $(N - 1)/e$; so q^n divides e, and a fortiori, q^n divides $p - 1$. □

The above statement looks more like a result on factors than a primality test. However, if it may be verified that each prime factor $p = mq^n + 1$ is greater than $\sqrt{N}$, then N is a prime. When q^n is fairly large, this verification is not too time consuming.

Pocklington gave also the following refinement of his result above:

Let $N - 1 = FR$, where $gcd(F, R) = 1$ and the factorization of F is known. Assume that for every prime q dividing F there exists an integer $a = a(q) > 1$ such that

(i) $a^{N-1} \equiv 1 \pmod{N}$,

(ii) $gcd(a^{(N-1)/q} - 1, N) = 1$.

Then each prime factor of N is of the form $mF + 1$, with $m \geq 1$.

The same comments apply here. So, if $F > \sqrt{N}$, then N is a prime.

This result is very useful to prove the primality of numbers of certain special form. The old criterion of Proth (1878) is easily deduced:

Test 4. Let $N = 2^n h + 1$ with h odd and $2^n > h$. Assume that there exists an integer $a > 1$ such that $a^{(N-1)/2} \equiv -1 \pmod{N}$. Then N is prime.

Proof. $N - 1 = 2^n h$, with h odd and $a^{N-1} \equiv 1 \pmod{N}$. Since N is odd, then $gcd(a^{(N-1)/2} - 1, N) = 1$. By the above result, each prime factor p of N is of the form $p = 2^n m + 1 > 2^n$. But $N = 2^n h + 1 < 2^{2n}$, hence $\sqrt{N} < 2^n < p$ and so N is prime.   □

In the following test (using the same notation) it is required to know that R (the nonfactored part of $N - 1$) has no prime factor less than a given bound B. Precisely:

Test 5. Let $N - 1 = FR$, where $gcd(F, R) = 1$, the factorization of F is known, B is such that $FB > \sqrt{N}$, and R has no prime factors less than B. Assume:

(i) For each prime q dividing F there exists an integer $a = a(q) > 1$ such that $a^{N-1} \equiv 1 \pmod{N}$ and $gcd(a^{(N-1)/q} - 1, N) = 1$.

(ii) There exists an integer $b > 1$ such that $b^{N-1} \equiv 1 \pmod{N}$ and $gcd(b^F - 1, N) = 1$.

Then N is a prime.

Proof. Let p be any prime factor of N, let e be the order of b modulo N, so e divides $p - 1$ and also e divides $N - 1 = FR$. Since e does not divide F, then $\gcd(e, R) \neq 1$, so there exists a prime q such that $q|e$, $q|R$; hence, $q|p - 1$. But, by the previous result of Pocklington, F divides $p - 1$; since $\gcd(F, R) = 1$, then qF divides $p - 1$. So $p - 1 \geq qF \geq BF > \sqrt{N}$. This implies that $p = N$, so N is a prime.   □

The paper of Brillhart, Lehmer & Selfridge (1975) contains other variants of these tests, which have been put to good use to determine the primality of numbers of the form $2^r + 1, 2^{2r} \pm 2^r + 1, 2^{2r-1} \pm 2^r + 1$.

I have already said enough and will make only one comment: these tests require prime factors of $N - 1$. Later, using linear recurring sequences, other tests will be presented, requiring prime factors of $N + 1$.

IV. Lucas Sequences

Let P, Q be nonzero integers.

Consider the polynomial $X^2 - PX + Q$; its discriminant is $D = P^2 - 4Q$ and the roots are

$$\left.\begin{array}{c}\alpha \\ \beta\end{array}\right\} = \frac{P \pm \sqrt{D}}{2}.$$

So

$$\begin{cases}\alpha + \beta = P, \\ \alpha\beta = Q, \\ \alpha - \beta = \sqrt{D}\end{cases}.$$

I shall assume that $D \neq 0$. Note that $D \equiv 0 \pmod 4$ or $D \equiv 1 \pmod 4$. Define the sequences of numbers

$$U_n(P, Q) = \frac{\alpha^n - \beta^n}{\alpha - \beta} \quad \text{and} \quad V_n(P, Q) = \alpha^n + \beta^n, \qquad \text{for } n \geq 0.$$

In particular, $U_0(P, Q) = 0$, $U_1(P, Q) = 1$, while $V_0(P, Q) = 2$, $V_1(P, Q) = P$.

The sequences

$$U(P, Q) = (U_n(P, Q))_{n \geq 1} \text{and} V(P, Q) = (V_n(P, Q))_{n \geq 1}$$

are called the *Lucas sequences associated to the pair* (P, Q). Special cases had been considered by Fibonacci, Fermat, and Pell, among others. Many particular facts were known about these sequences; however, the general theory was first developed by Lucas in a seminal paper, which appeared in Volume I of the *American Journal of Mathematics*, 1878. It is a long memoir with a rich content, relating Lucas sequences to many interesting topics, like trigonometric functions, continued fractions, the number of divisions in the algorithm of the greatest common divisor, and, also, primality tests. It is for this latter reason that I discuss Lucas sequences. If you are curious about the other connections that I have mentioned, look at the references at the end of the book and/or consult the paper in the library.

I should, however, warn that despite the importance of the paper, the methods employed are often indirect and cumbersome, so it is advisable to read Carmichael's long article of 1913, where he corrected errors and generalized results.

The first thing to note is that, for every $n \geq 2$,

$$\begin{aligned} U_n(P,Q) &= PU_{n-1}(P,Q) - QU_{n-2}(P,Q), \\ V_n(P,Q) &= PV_{n-1}(P,Q) - QV_{n-2}(P,Q) \end{aligned}$$

(just check it). So, these sequences deserve to be called *linear recurring sequences of order* 2 (each term depends linearly on the two preceding terms). Conversely, if P, Q are as indicated, and $D = P^2 - 4Q \neq 0$, if $W_0 = 0$ (resp. 2), $W_1 = 1$ (resp. P), if $W_n = PW_{n-1} - QW_{n-2}$ for $n \geq 2$, then Binet showed (in 1843) that

$$W_n = \frac{\alpha^n - \beta^n}{\alpha - \beta} \qquad (\text{resp.,} \ W_n = \alpha^n + \beta^n) \qquad \text{for } n \geq 0;$$

here α, β are the roots of the polynomial $X^2 - PX + Q$. This is trivial, because the sequences of numbers

$$(W_n)_{n \geq 0} \quad \text{and} \quad \left(\frac{\alpha^n - \beta^n}{\alpha - \beta} \right)_{n \geq 0} \qquad (\text{resp.,} \ (\alpha^n + \beta^n)_{n \geq 0}),$$

have the first two terms equal and both have the same linear second-order recurrence definition.

Before I continue, here are the main special cases that had been considered before the full theory was developed.

The sequence corresponding to $P = 1$, $Q = -1$, $U_0 = U_0(1, -1) = 0$, and $U_1 = U_1(1, -1) = 1$ was first considered by Fibonacci, and it begins as follows:

$$0 \quad 1 \quad 1 \quad 2 \quad 3 \quad 5 \quad 8 \quad 13 \quad 21 \quad 34 \quad 55 \quad 89 \quad 144 \quad 233$$

$$377 \quad 610 \quad 987 \quad 1597 \quad 2584 \quad 4181 \quad 6765 \quad \ldots.$$

These numbers appeared in a problem of Fibonacci concerning the number of offspring of rabbits; this is explained in all elementary books about these numbers. I do not care for such an explanation. As regards rabbits, I rather prefer to eat a good plate of "lapin chasseur" with fresh noodles.

The companion sequence of Fibonacci numbers, still with $P = 1$, $Q = -1$, is the sequence of Lucas numbers: $V_0 = V_0(1, -1) = 2$, $V_1 = V_1(1, -1) = 1$, and it begins as follows:

$$2 \quad 1 \quad 3 \quad 4 \quad 7 \quad 11 \quad 18 \quad 29 \quad 47 \quad 76 \quad 123 \quad 199 \quad 322$$

$$521 \quad 843 \quad 1364 \quad 2207 \quad 3571 \quad 5778 \quad 9349 \quad 15127 \quad \ldots.$$

If $P = 3$, $Q = 2$, then the sequences obtained are

$$U_n(3, 2) = 2^n - 1 \quad \text{and} \quad V_n(3, 2) = 2^n + 1, \quad \text{for } n \geq 0.$$

These sequences were the reason of many sleepless nights for Fermat (see details in Sections VI and VII). The sequences associated to $P = 2$, $Q = -1$, are called the Pell sequences; they begin as follows:

$$U_n(2, -1) : \quad 0 \quad 1 \quad 2 \quad 5 \quad 12 \quad 29 \quad 70 \quad 169 \quad 408$$
$$985 \quad 2378 \quad 5741 \quad 17223 \quad \ldots,$$
$$V_n(2, -1) : \quad 2 \quad 2 \quad 6 \quad 14 \quad 34 \quad 82 \quad 198 \quad 478 \quad 1154$$
$$2786 \quad 6726 \quad 16238 \quad 39202 \quad \ldots.$$

Lucas noted a great similarity between the sequences of numbers $U_n(P, Q)$ (resp. $V_n(P, Q)$) and $(a^n - b^n)/(a - b)$ (resp. $a^n + b^n$), where a, b are given, $a > b \geq 1$, $\gcd(a, b) = 1$ and $n \geq 0$. No wonder, one is a special case of the other. Just observe that for the pair $(a + b, ab)$, $D = (a - b)^2 \neq 0$, $\alpha = a$, $\beta = b$, so

$$U_n(a + b, ab) = \frac{a^n - b^n}{a - b}, \qquad V_n(a + b, ab) = a^n + b^n.$$

It is clearly desirable to extend the main results about the sequence of numbers $(a^n - b^n)/(a - b)$, $a^n + b^n$ (in what relates to divisibility and primality) for the wider class of Lucas sequences.

I shall therefore present the generalizations of Fermat's little theorem, Euler's theorem, etc., to Lucas sequences. There is no essential difficulty, but the development requires a surprising number of steps – true enough, all at an elementary level. In what follows, I shall record, one after the other, the facts needed to prove the main results. If you wish, work out the details. But I'm also explicitly giving the beginning of several Lucas sequences, so you may be happy just to check my statements numerically (see tables at the end of the section).

First, the algebraic facts, then the divisibility facts. To simplify the notations, I write only $U_n = U_n(P, Q)$, $V_n = V_n(P, Q)$.

I repeat:

(IV.1) $U_n = PU_{n-1} - QU_{n-2}(n \geq 2), \quad U_0 = 0, \quad U_1 = 1,$

$$V_n = PV_{n-1} - QV_{n-2}(n \geq 2), \quad V_0 = 2, \quad V_1 = P.$$

(IV.2) $U_{2n} = U_n V_n,$

$$V_{2n} = V_n^2 - 2Q^n.$$

(IV.3) $U_{m+n} = U_m V_n - Q^n U_{m-n},$

$$V_{m+n} = V_m V_n - Q^n V_{m-n}(\text{for } m \geq n).$$

(IV.4) $U_{m+n} = U_m U_{n+1} - QU_{m-1}U_n,$

$$2V_{m+n} = V_m V_n + DU_m U_n.$$

(IV.5) $DU_n = 2V_{n+1} - PV_n,$

$$V_n = 2U_{n+1} - PU_n.$$

(IV.6) $$U_n^2 = U_{n-1}U_{n+1} + Q^{n-1},$$

$$V_n^2 = DU_n^2 + 4Q^n.$$

(IV.7) $$U_mV_n - U_nV_m = 2Q^nU_{m-n}(\text{for } m \geq n),$$

$$U_mV_n + U_nV_m = 2U_{m+n}.$$

(IV.8) $$2^{n-1}U_n = \binom{n}{1}P^{n-1} + \binom{n}{3}P^{n-3}D + \binom{n}{5}P^{n-5}D^2 + \cdots,$$

$$2^{n-1}V_n = P^n + \binom{n}{2}P^{n-2}D + \binom{n}{4}P^{n-4}D^2 + \cdots.$$

(IV.9)   If m is odd and $k \geq 1$, then

$$
\begin{aligned}
D^{(m-1)/2}U_k^m &= U_{km} - \binom{m}{1}Q^kU_{k(m-2)} + \binom{m}{2}Q^{2k}U_{k(m-4)} \\
&\quad - \cdots \pm \binom{m}{\frac{m-1}{2}}Q^{\frac{m-1}{2}k}U_k,
\end{aligned}
$$

$$
\begin{aligned}
V_k^m &= V_{km} + \binom{m}{1}Q^kV_{k(m-2)} + \binom{m}{2}Q^{2k}V_{k(m-4)} \\
&\quad + \cdots + \binom{m}{\frac{m-1}{2}}Q^{\frac{m-1}{2}k}V_k.
\end{aligned}
$$

If m is even and $k \geq 1$, then

$$
\begin{aligned}
D^{m/2}U_k^m &= V_{km} - \binom{m}{1}Q^kV_{k(m-2)} + \binom{m}{2}Q^{2k}V_{k(m-4)} \\
&\quad + \cdots + \binom{m}{\frac{m}{2}}Q^{\frac{m}{2}k} \times 2,
\end{aligned}
$$

$$V_k^m = V_{km} + \binom{m}{1} Q^k V_{k(m-2)} + \binom{m}{2} Q^{2k} V_{k(m-4)}$$

$$+ \cdots + \binom{m}{\frac{m}{2}} Q^{\frac{m}{2}k} \times 2.$$

(IV.10) $U_m = V_{m-1} + QV_{m-3} + Q^2 V_{m-5} + \cdots + \text{(last summand)},$

where

$$\text{last summand} = \begin{cases} Q^{\frac{m-2}{2}} P & if\ m\ is\ even, \\ Q^{\frac{m-1}{2}} & if\ m\ is\ odd. \end{cases}$$

$$P^m = V_m + \binom{m}{1} QV_{m-2} + \binom{m}{2} Q^2 V_{m-4} + \cdots + \text{(last summand)},$$

where

$$\text{last summand} = \begin{cases} \binom{m}{\frac{m}{2}} Q^{\frac{m}{2}} & \text{if } m \text{ is even,} \\ \binom{m}{\frac{m-1}{2}} Q^{\frac{m-1}{2}} P & \text{if } m \text{ is odd.} \end{cases}$$

The following identity of Lagrange, dating from 1741, is required for the next property:

$$X^n + Y^n = (X+Y)^n - \frac{n}{1} XY(X+Y)^{n-2}$$

$$+ \frac{n}{2} \binom{n-3}{1} X^2 Y^2 (X+Y)^{n-4}$$

$$- \frac{n}{3} \binom{n-4}{2} X^3 Y^3 (X+Y)^{n-6} + \cdots$$

$$+ (-1)^r \frac{n}{r} \binom{n-r-1}{r-1} X^r Y^r (X+Y)^{n-2r}$$

$$\pm \cdots,$$

where the sum is extended for $2r \leq n$. Note that each coefficient is an integer.

(IV.11) If $m \geq 1$ and q is odd,

$$U_{mq} = D^{(q-1)/2} U_m^q + \frac{q}{1} Q^m D^{(q-3)/2} U_m^{q-2} + \frac{q}{2} \binom{q-3}{1} Q^{2m} D^{(q-5)/2} U_m^{q-4}$$

$$+ \cdots + \frac{q}{r}\binom{q-r-1}{r-1} Q^{mr} D^{(q-2r-1)/2} U_m^{q-2r} + \cdots + \text{(last summand)},$$

where the last summand is

$$\frac{q}{(q-1)/2}\binom{\frac{q-1}{2}}{\frac{q-3}{2}} Q^{\frac{q-1}{2}-m} U_m = qQ^{\frac{q-1}{2}-m} U_m.$$

Now, I begin to indicate, one after the other, the divisibility properties, in the order in which they may be proved.

(IV.12) $$U_n \equiv V_{n-1} \pmod{Q}$$

$$V_n \equiv P^n \pmod{Q}.$$

Hint: use (IV.10).

(IV.13) Let p be an odd prime, then

$$U_{kp} \equiv D^{\frac{p-1}{2}} U_k \pmod{p}$$

and, for $e \geq 1$,

$$U_{p^e} \equiv D^{\frac{p-1}{2}e} \pmod{p}.$$

In particular,

$$U_p \equiv \left(\frac{D}{p}\right) \pmod{p}.$$

Hint: Use (IV.9).

(IV.14) $$V_p \equiv P \pmod{p}.$$

Hint: use (IV.10).

(IV.15) If $n, k \geq 1$, then U_n divides U_{kn}.

Hint: use (IV.3).

(IV.16) If $n, k \geq 1$ and k is odd, then V_n divides V_{kn}.
Hint: use (IV.9).

Notation. If $n \geq 2$ and if there exists $r \geq 1$ such that n divides U_r, denote by $\rho(n) = \rho(n, U)$ the smallest such r.

(IV.17) Assume that $\rho(n)$ exists and $\gcd(n, 2Q) = 1$. Then $n|U_k$ if and only if $\rho(n)|k$.

Hint: use (IV.15) and (IV.7).

It will be seen that $\rho(n)$ exists, for many—not for all—values of n, such that $\gcd(n, 2Q) = 1$.

(IV.18) If Q is even and P is even, then U_n is even (for $n \geq 2$) and V_n is even (for $n \geq 1$).

If Q is even and P is odd, then U_n, V_n are odd (for $n \geq 1$).

If Q is odd and P is even, then $U_n \equiv n \pmod 2$ and V_n is even.

If Q is odd and P is odd, then U_n, V_n are even if 3 divides n, while U_n, V_n are odd, otherwise.

In particular, if U_n is even, then V_n is even.

Hint: use (IV.12), (IV.5), (IV.2), (IV.6), and (IV.1).

Here is the first main result, which is a companion of (IV.18) and generalizes Fermat's little theorem:

(IV.19) Let p be an odd prime.

If $p|P$ and $p|Q$, then $p|U_k$ for every $k > 1$.

If $p|P$ and $p \nmid Q$, then $p|U_k$ exactly when k is even.

If $p \nmid P$ and $p|Q$, then $p \nmid U_n$ for every $n \geq 1$.

If $p \nmid P$, $p \nmid Q$, and $p|D$, then $p|U_k$ exactly when $p|k$.

If $p \nmid PQD$, then $p|U_{\psi(p)}$, where $\psi(p) = p - (D/p)$, and (D/p) denotes the Legendre symbol.

Proof. If $p|P$ and $p|Q$, by (IV.1) $p|U_k$ for every $k > 1$.

If $p|P = U_2$, by (IV.15) $p|U_{2k}$ for every $k \geq 1$. Since $p \nmid Q$, and $U_{2k+1} = PU_{2k} - QU_{2k-1}$, by induction, $p \nmid U_{2k+1}$.

If $p \nmid P$ and $p|Q$, by induction and (IV.1), $p \nmid U_n$ for every $n \geq 1$.

If $p \nmid PQ$ and $p|D$, by (IV.8), $2^{p-1}U_p \equiv 0 \pmod p$ so $p|U_p$. On the other hand, if $p \nmid n$, then by (IV.8) $2^{n-1}U_n \equiv nP^{n-1} \not\equiv 0 \pmod p$, so $p \nmid U_n$.

Finally the more interesting case: assume $p \nmid PQD$.

If $(D/p) = -1$, then by (IV.8)

$$2^p U_{p+1} = \binom{p+1}{1}P^p + \binom{p+1}{3}P^{p-2}D + \cdots + \binom{p+1}{p}PD^{\frac{p-1}{2}}$$

$$\equiv P + PD^{\frac{p-1}{2}} \equiv 0 \pmod p, \quad \text{so } p|U_{p+1}.$$

If $(D/p) = 1$, there exists C such that $P^2 - 4Q = D \equiv C^2$ (mod p); hence, $P^2 \not\equiv C^2$ (mod p) and $p \nmid C$. By (IV.8), noting that

$$\binom{p-1}{1} \equiv -1 \pmod{p}, \quad \binom{p-1}{3} \equiv -1 \pmod{p}, \dots:$$

$$2^{p-2}U_{p-1} = \binom{p-1}{1}P^{p-2} + \binom{p-1}{3}P^{p-4}D + \binom{p-1}{5}P^{p-6}D^2 + \cdots +$$

$$\binom{p-1}{p-2}PD^{\frac{p-3}{2}} \equiv -[P^{p-2} + P^{p-4}D + P^{p-6}D^2 + \cdots + PD^{\frac{p-3}{2}}]$$

$$\equiv -P\left(\frac{P^{p-1} - D^{\frac{p-1}{2}}}{P^2 - D}\right) \equiv -P\frac{P^{p-1} - C^{p-1}}{P^2 - C^2} \equiv 0 \pmod{p}.$$

So $p|U_{p-1}$.   □

If I want to use the notation $\rho(p)$ introduced before, some of the assertions of (IV.19) may be restated as follows:

If p is an odd prime and $p \nmid Q$:

If $p|P$, then $\rho(p) = 2$.

If $p \nmid P$, $p|D$, then $\rho(p) = p$.

If $p \nmid PD$, then $\rho(p)$ divides $\psi(p)$.

Don't conclude hastily that, in this latter case, $\rho(p) = \psi(p)$. I shall return to this point, after I list the main properties of the Lucas sequences.

For the special Lucas sequence $U_n(a + 1, a)$, the discriminant is $D = (a - 1)^2$; so if $p \nmid a(a^2 - 1)$, then

$$\left(\frac{D}{p}\right) = 1 \quad \text{and} \quad p|U_{p-1} = \frac{a^{p-1} - 1}{a - 1},$$

so $p|a^{p-1} - 1$ (this is trivial if $p|a^2 - 1$)—which is Fermat's little theorem.

(IV.20) Let $e \geq 1$, and let p^e be the exact power of p dividing U_m. If $p \nmid k$ and $f \geq 1$, then p^{e+f} divides U_{mkp^f}.

Moreover, if $p|Q$ and $p^e \neq 2$, then p^{e+f} is the exact power of p dividing U_{mkp^f}, while if $p^e = 2$ then $U_{mk}/2$ is odd.

Hint: use (IV.19), (IV.18), (IV.11), and (IV.6).

And now the generalization of Euler's theorem:

If α, β are roots of $X^2 - PX + Q$, define the symbol:

$$\left(\frac{\alpha, \beta}{2}\right) = \begin{cases} 1 & \text{if } Q \text{ is even,} \\ 0 & \text{if } Q \text{ is odd, } P \text{ even,} \\ -1 & \text{if } Q \text{ is odd, } P \text{ odd.} \end{cases}$$

and $p \neq 2$:

$$\left(\frac{\alpha, \beta}{p}\right) = \left(\frac{D}{p}\right)$$

(so it is 0 if $p|D$). Put

$$\psi_{\alpha, \beta}(p) = p - \left(\frac{\alpha, \beta}{p}\right)$$

for every prime p, also

$$\psi_{\alpha, \beta}(p^e) = p^{e-1} \psi_{\alpha, \beta}(p) \text{ for } e \geq 1.$$

If $n = \prod_{p|n} p^e$, define the Carmichael function

$$\lambda_{\alpha, \beta}(n) = \ell\text{cm}\{\psi_{\alpha, \beta}(p^e)\}$$

(where ℓcm denotes the least common multiple), and define the generalized Euler function

$$\psi_{\alpha, \beta}(n) = \prod_{p|n} \psi_{\alpha, \beta}(p^e).$$

So $\lambda_{\alpha, \beta}(n)$ divides $\psi_{\alpha, \beta}(n)$.

It is easy to check that $\psi_{a,1}(p) = p-1 = \phi(p)$ for every prime p not dividing a; so if $\gcd(a, n) = 1$, then $\psi_{a,1}(n) = \phi(n)$ and also $\lambda_{a,1}(n) = \lambda(n)$, where $\lambda(n)$ is the function, also defined by Carmichael, and considered in Section II.

And here is the extension of Euler's theorem:

(IV.21) If $\gcd(n, Q) = 1$, then n divides $U_{\lambda_{\alpha, \beta}(n)}$; hence, also n divides $U_{\psi_{\alpha, \beta}(n)}$.

Hint: use (IV.19) and (IV.20).

It should be said that the divisibility properties of the companion sequence$(V_n)_{n \geq 1}$ are not so simple to describe. Note, for example,

(IV.22) If $p \nmid 2QD$, then $V_{p-(D/p)} \equiv 2Q^{\frac{1}{2}[1-(D/p)]} \pmod{p}$.

Hint: use (IV.5), (IV.13), (IV.19), and (IV.14).

This may be applied to give divisibility results for $U_{\psi(p)/2}$ and $V_{\psi(p)/2}$.

(IV.23) Assume that $p \nmid 2QD$. Then
$p|U_{\psi(p)/2}$ if and only if $(Q/p) = 1$;
$p|V_{\psi(p)/2}$ if and only if $(Q/p) = -1$.
Hint: for the first assertion, use (IV.2), (IV.6), (IV.22) and the congruence $(Q/p) \equiv Q^{\frac{p-1}{2}} \pmod{p}$.

For the second assertion, use (IV.2), (IV.19), the first assertion, and also (IV.6).

For the next results, I shall assume that $\gcd(P, Q) = 1$.

(IV.24) $\gcd(U_n, Q) = 1$, $\gcd(V_n, Q) = 1$.
Hint: use (IV.12).

(IV.25) $\gcd(U_n, V_n) = 1$ or 2.
Hint: use (IV.16), and (IV.24).

(IV.26) If $d = \gcd(m, n)$, then $U_d = \gcd(U_m, U_n)$.
Hint: use (IV.15), (IV.7), (IV.24), (IV.18), and (IV.6).

This proof is actually not so easy, and requires the use of the Lucas sequence $(U_n(V_d, Q^d))_{n \geq 0}$.

(IV.27) If $\gcd(m, n) = 1$, then $\gcd(U_m, U_n) = 1$.
No hint for this one.

(IV.28) If $d = \gcd(m, n)$ and m/d, n/d are odd, then $V_d = \gcd(V_m, V_n)$.
Hint: use the same proof as for (IV.26).

And here is a result similar to (IV.17), but with the assumption that $\gcd(P, Q) = 1$:

(IV.29) Assume that $\rho(n)$ exists. Then $n|U_k$ if and only if $\rho(n)|k$.
Hint: use (IV.15), (IV.24), and (IV.3).

I pause to write explicitly what happens for the Fibonacci numbers U_n and Lucas numbers V_n; now $P = 1$, $Q = -1$, $D = 5$.

Property (IV.18) becomes the *law of appearance* of p; even though I am writing this text on Halloween's evening, it would hurt me to call it the "apparition law" (as it was badly translated from the French *loi d'apparition*; in all English dictionaries "apparition" means "ghost"). Law of apparition (oops!, appearance) of p:

$p|U_{p-1}$ if $(5/p) = 1$, that is, $p \equiv \pm 1 \pmod{10}$
$p|U_{p+1}$ if $(5/p) = -1$, that is, $p \equiv \pm 3 \pmod{10}$.

Property (IV.19) is the *law of repetition*.

For the Lucas numbers, the following properties hold:

$p|V_{p-1} - 2$ if $(5/p) = 1$, that is, $p \equiv \pm 1 \pmod{10}$
$p|V_{p+1} + 2$ if $(5/p) = -1$, that is, $p \equiv \pm 3 \pmod{10}$.

Jarden showed in 1958 that, for the Fibonacci sequence, the function

$$\frac{\psi(p)}{\rho(p)} = \frac{p - (5/p)}{\rho(p)}$$

is unbounded (when the prime p tends to infinity).

This result was generalized by Kiss & Phong in 1978: there exists $C > 0$ (depending only on P, Q) such that $\psi(p)/\rho(p)$ is unbounded, but still $\psi(p)/\rho(p) < C[p/(\log p)]$ (when the prime p tends to infinity).

Now I shall indicate the behaviour of Lucas sequences modulo a prime p.

If $p = 2$, this is as described in (IV.18). For example, if P, Q are odd, then the sequences $U_n \bmod 2$, $V_n \bmod 2$ are equal to

$$1 \quad 1 \quad 0 \quad 1 \quad 1 \quad 0 \quad 1 \quad 1 \quad 0 \quad \ldots .$$

It is more interesting when p is an odd prime.

(IV.30) If $p \nmid 2QD$ and $(D/p) = 1$, then

$$U_{n+p-1} \equiv U_n \pmod{p},$$
$$V_{n+p-1} \equiv V_n \pmod{p}.$$

Thus, the sequences $U_n \bmod p$, $V_n \bmod p$ have period $p - 1$.

Proof. By (IV.4), $U_{n+p-1} = U_n U_p - Q U_{n-1} U_{p-1}$; by (IV.19), $\rho(p)$ divides $p - (D/p) = p - 1$; by (IV.15), $p|U_{p-1}$; this is also true if $p|P$, $p \nmid Q$, because then $p - 1$ is even, so $p|U_{p-1}$, by (IV.19). By (IV.13),

$$U_p \equiv (D/p) \equiv 1 \pmod{p}.$$

So $U_{n+p-1} \equiv U_n \pmod{p}$.

Now, by (IV.5), $V_{n+p-1} = 2U_{n+p} - PU_n \equiv 2U_{n+1} - PU_n \equiv V_n \pmod{p}$.   □

The companion result is the following:

(IV.31) Let $p \nmid 2QD$, let e be the order of Q mod p. If $(D/p) = -1$, then

$$U_{n+e(p+1)} \equiv U_n \pmod{p},$$
$$V_{n+e(p+1)} \equiv V_n \pmod{p}.$$

Thus, the sequences U_n mod p, V_n mod p have period $e(p+1)$.

Proof. By (IV.19), (IV.15),

$$p \mid U_{p-(D/p)} = U_{p+1}.$$

By (IV.22), $V_{p+1} \equiv 2Q \pmod{p}$. Now I show, by induction on $r \geq 1$, that $V_{r(p+1)} \equiv 2Q^r \pmod{p}$.

If this is true for $r \geq 1$, then by (IV.4)

$$2V_{(r+1)(p+1)} = V_{r(p+1)}V_{p+1} + DU_{r(p+1)}U_{p+1} \equiv 4Q^{r+1} \pmod{p},$$

so $V_{(r+1)(p+1)} \equiv 2Q^{r+1} \pmod{p}$. In particular, $V_{e(p+1)} \equiv 2Q^e \equiv 2 \pmod{p}$.

By (IV.7),

$$U_{n+e(p+1)}V_{e(p+1)} - U_{e(p+1)}V_{n+e(p+1)} = 2Q^{e(p+1)}U_n,$$

hence $2U_{n+e(p+1)} \equiv 2U_n \pmod{p}$ and the first congruence is established.

The second congruence follows using (IV.5).   □

It is good to summarize some of the preceding results, by writing explicitly the sets

$$\mathcal{P}(U) = \{p \text{ prime} \mid \text{there exists } n \text{ such that } p \mid U_n\},$$
$$\mathcal{P}(V) = \{p \text{ prime} \mid \text{there exists } n \text{ such that } p \mid V_n\}.$$

These are the prime divisors of the sequence $U = (U_n)_{n \geq 1}$, $V = (V_n)_{n \geq 1}$.

The parameters (P, Q) are assumed to be nonzero relatively prime integers and the discriminant is $D = P^2 - 4Q \neq 0$.

A first case arises if there exists $n > 1$ such that $U_n = 0$; equivalently, $\alpha^n = \beta^n$, that is α/β is a root of unity. If n is the smallest such index, then $U_r \neq 0$ for $r = 1, \ldots, n - 1$ and $U_{nk+r} = \alpha^{nk} U_r$ (for every $k \geq 1$), so $\mathcal{P}(U)$ consists of the prime divisors of $U_2 \cdots U_{n-1}$. Similarly, $\mathcal{P}(V)$ consists of the same prime numbers.

The more interesting case is when α/β is not a root of unity, so $U_n \neq 0$, $V_n \neq 0$ for every $n \geq 1$. Then $\mathcal{P}(U) = \{ p \text{ prime} \mid p \text{ does not divide } Q \}$.

This follows from (IV.18) and (IV.19). In particular, for the sequence of Fibonacci numbers, $\mathcal{P}(U)$ is the set of all primes.

Nothing so precise may be said about the companion Lucas sequence $V = (V_n)_{n \geq 1}$. From $U_{2n} = U_n V_n$ ($n \geq 1$) it follows that $\mathcal{P}(V)$ is a subset of $\mathcal{P}(U)$. From (IV.18), $2 \in \mathcal{P}(V)$ if and only if Q is odd. Also, from (IV.24) and (IV.6), if $p \neq 2$ and if $p|DQ$, then $p \notin \mathcal{P}(V)$, while if $p \nmid 2DQ$ and $(Q/p) = -1$, then $p \in \mathcal{P}(V)$ [see (IV.23)]; on the other hand, if $p \nmid 2DQ$, $(Q/p) = 1$, and $(D/p) = -(-1/p)$, then $p \notin \mathcal{P}(V)$. This does not determine, without a further analysis, whether a prime p, such that $p \nmid 2DQ$, $(Q/p) = 1$, and $(D/p) = (-1/p)$ belongs, or does not belong, to $\mathcal{P}(V)$.

At any rate, it shows that $\mathcal{P}(V)$ is also an infinite set.

For the sequence of Lucas numbers, with $P = 1$, $Q = -1$, $D = 5$, the preceding facts may be explicitly stated as follows:

if $p \equiv 3, 7, 11, 19 \pmod{20}$, then $p \in \mathcal{P}(V)$;
if $p \equiv 13, 17 \pmod{20}$, then $p \notin \mathcal{P}(V)$.

For $p \equiv 1, 9 \pmod{20}$, no decision may be obtained without a careful study, as, for example, that done by Ward in 1961. Already in 1958 Jarden had shown that there exist infinitely many primes p, $p \equiv 1 \pmod{20}$, such that $p \notin \mathcal{P}(V)$, and, on the other hand, there exist also infinitely many primes p, $p \equiv 1 \pmod{40}$, such that $p \in \mathcal{P}(V)$.

Later, in Chapter 5, Section VIII, I shall return to the study of the sets $\mathcal{P}(U)$, $\mathcal{P}(V)$, asking for their density in the set of all primes.

In analogy with the theorem of Bang and Zsigmondy, Carmichael also considered the primitive prime factors of the Lucas sequences, with parameters (P, Q): p is a primitive prime factor of U_k (resp. V_k) if $p|U_k$ (resp. $p|V_k$), but p does not divide any preceding number in the sequence in question.

The proof of Zsigmondy's theorem is not too simple; here it is somewhat more delicate.

Carmichael showed that if the discriminant D is positive, then for every $n \neq 1, 2, 6$, U_n has a primitive prime factor, except if $n = 12$

and $P = \pm 1$, $Q = -1$.

Moreover, if D is a square, then it is better: for every n, U_n has a primitive prime factor, except if $n = 6$, $P = \pm 3$, $Q = 2$.

Do you recognize that this second statement includes Zsigmondy's theorem? Also, if $P = 1$, $Q = -1$ the exception is the Fibonacci number $U_{12} = 144$.

For the companion sequence, if $D > 0$, then for every $n \neq 1, 3$, V_n has a primitive prime factor, except if $n = 6$, $P = \pm 1$, $Q = -1$ (the Lucas number $V_6 = 18$). Moreover, if D is a square, then the only exception is $n = 3$, $P = \pm 3$, $Q = 2$, also contained in Zsigmondy's theorem.

If, however, $D < 0$, the result indicated is no longer true. Thus, as Carmichael already noted, if $P = 1$, $Q = 2$, then for $n = 1, 2, 3, 5, 8$, $12, 13, 18$, U_n has no primitive prime factors.

Schinzel showed the following in 1962:

Let $(U_n)_{n \geq 0}$ be the Lucas sequence with relatively prime parameters (P, Q) and assume that the discriminant is $D < 0$. Assume that α/β is not a root of unity. Then there exists n_0 (depending on P, Q), effectively computable, such that if $n > n_0$, then U_n has a primitive prime factor.

Later, in 1974, Schinzel proved the same result with an absolute constant n_0—independent of the Lucas sequence.

Making use of the methods of Baker, Stewart determined in 1977 that if $n > e^{452} 2^{67}$, then U_n has a primitive prime factor. Moreover, Stewart also showed that if n is given ($n \neq 6$, $n > 4$), there are only finitely many Lucas sequences, which may be determined explicitly (so says Stewart, without doing it), for which U_n has no primitive prime factor.

It is interesting to consider the primitive part U_n^* of U_n:

$$U_n = U_n^* U_n' \quad \text{with} \quad \gcd(U_n^*, U_n') = 1$$

and p divides U_n^* if and only if p is a primitive prime factor of U_n.

In 1963, Schinzel indicated conditions for the existence of two (or even $e > 2$) distinct primitive prime factors. It follows that if $D > 0$ or $D < 0$ and α/β is not a root of unity, there exist infinitely many n such that the primitive part U_n^* is composite.

Can one say anything about U_n^* being square-free? This is a very deep question. Just think of the special case where $P = 3$, $Q = 2$, which gives the sequence $2^n - 1$ (see my comments in Section II).

$$P = 1, Q = -1$$

Fibonacci Numbers		Lucas Numbers	
$U(0) = 0$	$U(1) = 1$	$V(0) = 2$	$V(1) = 1$
$U(2) = 1$		$V(2) = 3$	
$U(3) = 2$		$V(3) = 4$	
$U(4) = 3$		$V(4) = 7$	
$U(5) = 5$		$V(5) = 11$	
$U(6) = 8$		$V(6) = 18$	
$U(7) = 13$		$V(7) = 29$	
$U(8) = 21$		$V(8) = 47$	
$U(9) = 34$		$V(9) = 76$	
$U(10) = 55$		$V(10) = 123$	
$U(11) = 89$		$V(11) = 199$	
$U(12) = 144$		$V(12) = 322$	
$U(13) = 233$		$V(13) = 521$	
$U(14) = 377$		$V(14) = 843$	
$U(15) = 610$		$V(15) = 1364$	
$U(16) = 987$		$V(16) = 2207$	
$U(17) = 1597$		$V(17) = 3571$	
$U(18) = 2584$		$V(18) = 5778$	
$U(19) = 4181$		$V(19) = 9349$	
$U(20) = 6765$		$V(20) = 15127$	
$U(21) = 10946$		$V(21) = 24476$	
$U(22) = 17711$		$V(22) = 39603$	
$U(23) = 28657$		$V(23) = 64079$	
$U(24) = 46368$		$V(24) = 103682$	
$U(25) = 75025$		$V(25) = 167761$	
$U(26) = 121393$		$V(26) = 271443$	
$U(27) = 196418$		$V(27) = 439204$	
$U(28) = 317811$		$V(28) = 710647$	
$U(29) = 514229$		$V(29) = 1149851$	
$U(30) = 832040$		$V(30) = 1860498$	
$U(31) = 1346269$		$V(31) = 3010349$	
$U(32) = 2178309$		$V(32) = 4870847$	
$U(33) = 3524578$		$V(33) = 7881196$	
$U(34) = 5702887$		$V(34) = 12752043$	
$U(35) = 9227465$		$V(35) = 20633239$	
$U(36) = 14930352$		$V(36) = 33385282$	
$U(37) = 24157817$		$V(37) = 54018521$	
$U(38) = 39088169$		$V(38) = 87403803$	

$$P = 3, \; Q = 2$$

Numbers $2^n - 1$	Numbers $2^n + 1$
$U(0) = 0 \quad U(1) = 1$	$V(0) = 2 \quad V(1) = 3$
$U(2) = 3$	$V(2) = 5$
$U(3) = 7$	$V(3) = 9$
$U(4) = 15$	$V(4) = 17$
$U(5) = 31$	$V(5) = 33$
$U(6) = 63$	$V(6) = 65$
$U(7) = 127$	$V(7) = 129$
$U(8) = 255$	$V(8) = 257$
$U(9) = 511$	$V(9) = 513$
$U(10) = 1023$	$V(10) = 1025$
$U(11) = 2047$	$V(11) = 2049$
$U(12) = 4095$	$V(12) = 4097$
$U(13) = 8191$	$V(13) = 8193$
$U(14) = 16383$	$V(14) = 16385$
$U(15) = 32767$	$V(15) = 32769$
$U(16) = 65535$	$V(16) = 65537$
$U(17) = 131071$	$V(17) = 131073$
$U(18) = 262143$	$V(18) = 262145$
$U(19) = 524287$	$V(19) = 524289$
$U(20) = 1048575$	$V(20) = 1048577$
$U(21) = 2097151$	$V(21) = 2097153$
$U(22) = 4194303$	$V(22) = 4194305$
$U(23) = 8388607$	$V(23) = 8388609$
$U(24) = 16777215$	$V(24) = 16777217$
$U(25) = 33554431$	$V(25) = 33554433$
$U(26) = 67108863$	$V(26) = 67108865$
$U(27) = 134217727$	$V(27) = 134217729$
$U(28) = 268435455$	$V(28) = 268435457$
$U(29) = 536870911$	$V(29) = 536870913$
$U(30) = 1073741823$	$V(30) = 1073741825$
$U(31) = 2147483647$	$V(31) = 2147483649$
$U(32) = 4294967295$	$V(32) = 4294967297$
$U(33) = 8589934591$	$V(33) = 8589934593$
$U(34) = 17179869183$	$V(34) = 17179869185$
$U(35) = 34359738367$	$V(35) = 34359738369$
$U(36) = 68719476735$	$V(36) = 68719476637$
$U(37) = 137438953471$	$V(37) = 137438953473$
$U(38) = 274877906943$	$V(38) = 274877906945$

$$P = 2, \ Q = -1$$

Pell Numbers	Companion Pell Numbers
$U(0) = 0 \quad U(1) = 1$	$V(0) = 2 \quad V(1) = 2$
$U(2) = 2$	$V(2) = 6$
$U(3) = 5$	$V(3) = 14$
$U(4) = 12$	$V(4) = 34$
$U(5) = 29$	$V(5) = 82$
$U(6) = 70$	$V(6) = 198$
$U(7) = 169$	$V(7) = 478$
$U(8) = 408$	$V(8) = 1154$
$U(9) = 985$	$V(9) = 2786$
$U(10) = 2378$	$V(10) = 6726$
$U(11) = 5741$	$V(11) = 16238$
$U(12) = 13860$	$V(12) = 39202$
$U(13) = 33461$	$V(13) = 94642$
$U(14) = 80782$	$V(14) = 228486$
$U(15) = 195025$	$V(15) = 551614$
$U(16) = 470832$	$V(16) = 1331714$
$U(17) = 1136689$	$V(17) = 3215042$
$U(18) = 2744210$	$V(18) = 7761798$
$U(19) = 6625109$	$V(19) = 18738638$
$U(20) = 15994428$	$V(20) = 45239074$
$U(21) = 38613965$	$V(21) = 109216786$
$U(22) = 93222358$	$V(22) = 263672646$
$U(23) = 225058681$	$V(23) = 636562078$
$U(24) = 543339720$	$V(24) = 1536796802$
$U(25) = 1311738121$	$V(25) = 3710155682$
$U(26) = 3166815962$	$V(26) = 8957108166$
$U(27) = 7645370045$	$V(27) = 21624372014$
$U(28) = 18457556052$	$V(28) = 52205852194$
$U(29) = 44560482149$	$V(29) = 126036076402$
$U(30) = 107578520350$	$V(30) = 304278004998$
$U(31) = 259717522849$	$V(31) = 734592086398$
$U(32) = 627013566048$	$V(32) = 1773462177794$
$U(33) = 1513744654945$	$V(33) = 4281516441986$
$U(34) = 3654502875938$	$V(34) = 10336495061766$
$U(35) = 8822750406821$	$V(35) = 24954506565518$
$U(36) = 21300003689580$	$V(36) = 60245508192802$
$U(37) = 51422757785981$	$V(37) = 145445522951122$
$U(38) = 124145519261542$	$V(38) = 351136554095046$

$$P = 4,\ Q = 3$$

Numbers	Companion Numbers
$U(0) = 0 \quad U(1) = 1$	$V(0) = 2 \quad V(1) = 4$
$U(2) = 4$	$V(2) = 10$
$U(3) = 13$	$V(3) = 28$
$U(4) = 40$	$V(4) = 82$
$U(5) = 121$	$V(5) = 244$
$U(6) = 364$	$V(6) = 730$
$U(7) = 1093$	$V(7) = 2188$
$U(8) = 3280$	$V(8) = 6562$
$U(9) = 9841$	$V(9) = 19684$
$U(10) = 29524$	$V(10) = 59050$
$U(11) = 88573$	$V(11) = 177148$
$U(12) = 265720$	$V(12) = 531442$
$U(13) = 797161$	$V(13) = 1594324$
$U(14) = 2391484$	$V(14) = 4782970$
$U(15) = 7174453$	$V(15) = 14348908$
$U(16) = 21523360$	$V(16) = 43046722$
$U(17) = 64570081$	$V(17) = 129140164$
$U(18) = 193710244$	$V(18) = 387420490$
$U(19) = 581130733$	$V(19) = 1162261468$
$U(20) = 1743392200$	$V(20) = 3486784402$
$U(21) = 5230176601$	$V(21) = 10460353204$
$U(22) = 15690529804$	$V(22) = 31381059610$
$U(23) = 47071589413$	$V(23) = 94143178828$
$U(24) = 141214768240$	$V(24) = 282429536482$
$U(25) = 423644304721$	$V(25) = 847288609444$
$U(26) = 1270932914164$	$V(26) = 2541865828330$
$U(27) = 3812798742493$	$V(27) = 7625597484988$
$U(28) = 11438396227480$	$V(28) = 22876792454962$
$U(29) = 34315188682441$	$V(29) = 68630377364884$
$U(30) = 102945566047324$	$V(30) = 205891132094650$
$U(31) = 308836698141973$	$V(31) = 617673396283948$
$U(32) = 926510094425920$	$V(32) = 1853020188851842$
$U(33) = 2779530283277761$	$V(33) = 5559060566555524$

V. Primality Tests Based on Lucas Sequences

Lucas began, Lehmer continued, others refined. The primality tests of N, to be presented now, require the knowledge of prime factors of $N + 1$, and they complement the tests indicated in Section III, which needed the prime factors of $N - 1$. Now, the tool will be the Lucas sequences. By (IV.18), if N is an odd prime, if $U = (U_n)_{n \geq 0}$ is a Lucas sequence with discriminant D, and the Jacobi symbol $(D/N) = -1$, then N divides $U_{N-(D/N)} = U_{N+1}$.

However, I note right away (as I did in Section III) that a crude converse does not hold, because there exist composite integers N, and Lucas sequences $(U_n)_{n \geq 0}$ with discriminant D, such that N divides $U_{N-(D/N)}$. Such numbers will be studied in Section X.

It will be convenient to introduce for every integer $D > 1$ the function ψ_D, defined as follows:

If $N = \prod_{i=1}^{s} p_i^{e_i}$, let

$$\psi_D(N) = \frac{1}{2^{s-1}} \prod_{i=1}^{s} p_i^{e_i-1} \left(p_i - \left(\frac{D}{p_i} \right) \right).$$

Note that if $(U_n)_{n \geq 0}$ is a Lucas sequence with discriminant D, if α, β are the roots of the associated polynomial, then the function $\psi_{\alpha,\beta}$ considered in Section IV is related to ψ_D as follows: $\psi_{\alpha,\beta}(N) = 2^{s-1}\psi_D(N)$.

As it will be necessary to consider simultaneously several Lucas sequences with the same discriminant D, it is preferable to work with ψ_D, and not with the functions $\psi_{\alpha,\beta}$ corresponding to the various sequences.

Note, for example, that if $U(P,Q)$ has discriminant D, if $P' = P + 2$, $Q' = P + Q + 1$, then also $U(P',Q')$ has discriminant D.

It is good to start with some preparatory and easy results.

(V.1) If N is odd, $\gcd(N, D) = 1$, then $\psi_D(N) = N - (D/N)$ if and only if N is a prime.

Proof. If N is a prime, by definition $\psi_D(N) = N - (D/N)$.

If $N = p^e$ with p prime, $e \geq 2$, then $\psi_D(N)$ is a multiple of p, while $N - (D/N)$ is not.

If $N = \prod_{i=1}^{s} p_i^{e_i}$, with $s \geq 2$, then

$$\psi_D(N) \leq \frac{1}{2^s - 1} \prod_{i=1}^{e_i-1} p_i^{e_i-1}(p_i + 1)$$

$$= 2N \prod_{i=1} \frac{1}{2}\left(1 + \frac{1}{p_i}\right) \le 2N \times \frac{2}{3} \times \frac{3}{5} \times \cdots$$

$$\le \frac{4N}{5} < N - 1,$$

since $N > 5$. □

(V.2) If N is odd, $\gcd(N, D) = 1$, and $N - (D/N)$ divides $\psi_D(N)$, then N is a prime.

Proof. Assume that N is composite. First, let $N = p^e$, with p prime, $e \ge 2$; then $\psi_D(N) = p^e - p^{e-1}(D/p)$. Hence,

$$p^e - p^{e-1} < p^e - 1 \le p^e - (D/N) \le p^e - p^{e-1}(D/p),$$

so $(D/p) = -1$ and $p^e \pm 1$ divides $p^e + p^{e-1} < 2p^e - 2$, which is impossible.

If N has at least two distinct prime factors, it was seen in (V.1) that $\psi_D(N) < N - 1 \le N - (D/N)$, which is contrary to the hypothesis. So N must be a prime. □

(V.3) If N is odd, $U = U(P, Q)$ is a Lucas sequence with discriminant D, and $\gcd(N, QD) = 1$, then $N | U_{\psi_D(N)}$.

Proof. Since $\gcd(N, Q) = 1$, then by (IV.12) N divides $\lambda_{\alpha,\beta}(N)$, where α, β are the roots of $X^2 - PX + Q$. If $N = \prod_{i=1}^{S} p_i^{e_i}$, then

$$\lambda_{\alpha,\beta}(N) = \ell cm\left\{p_i^{e_i-1} - \left(p_i\left(\frac{D}{p_i}\right)\right)\right\} = 2\,\ell cm\left\{\frac{1}{2}p_i^{e_i-1}\left(p_i - \left(\frac{D}{p_i}\right)\right)\right\}$$

and $\lambda_{\alpha,\beta}(N)$ divides

$$2\prod_{i=1}^{s} \frac{1}{2}p_i^{e_i-1}\left(p_i - \left(\frac{D}{p_i}\right)\right) = \psi_D(N).$$

By (IV.15), N divides $U_{\psi_D(N)}$. □

(V.4) If N is odd, $U = U(P, Q)$ is a Lucas sequence with discriminant D such that $(D/N) = -1$, and N divides U_{N+1}, then $\gcd(N, QD) = 1$.

Proof. Since $(D/N) \neq 0$, then $\gcd(N, D) = 1$.

If there exists a prime p such that $p|N$ and $p|Q$, since $p \nmid D = P^2 - 4Q$, then $p \nmid P$. By (IV.18) $p \nmid U_n$ for every $n \geq 1$, which is contrary to the hypothesis. So $\gcd(N, Q) = 1$. $\square$

One more result which will be needed is the following:

(V.5) Let N be odd and q be any prime factor of $N + 1$. Assume that $U = U(P, Q)$ and $V = V(P, Q)$ are the Lucas sequences associated with the integers P, Q, having discriminant $D \neq 0$. Assume $\gcd(P, Q) = 1$ or $\gcd(N, Q) = 1$. If N divides $U_{(N+1)/q}$ and $V_{(N+1)/2}$ then N divides $V_{(N+1)/2q}$.

Proof.

$$\frac{N+1}{2} = \frac{N+1}{2q} + \frac{N+1}{q}u \quad \text{with} \quad u = \frac{q-1}{2}.$$

By (IV.4):

$$2V_{(N+1)/2} = V_{(N+1)/2q}V_{[(N+1)/q]u} + DU_{(N+1)/2q}U_{[(U+1)/q]u}.$$

By (IV.15), N divides $U_{[(N+1)/q]u}$ so N divides $V_{(N+1)/2q}V_{[(N+1)/q]u}$.
If $\gcd(P, Q) = 1$, by (IV.21) $\gcd(U_{[(N+1)/q]u}, V_{[(N+1)/q]u}) = 1$ or 2, hence $\gcd(N, V_{[(N+1)/q]u}) = 1$, so N divides $V_{(N+1)/2q}$.
If $\gcd(N, Q) = 1$ and if there exists a prime p dividing N and $V_{[(N+1)/q]u}$, then by (IV.6) p divides also $4Q$; since p is odd, then $p|Q$, which is a contradiction. $\square$

Before indicating primality tests, it is easy to give sufficient conditions for a number to be composite:

Let $N > 1$ be an odd integer. Assume that there exists a Lucas sequence $(U_n)_{n\geq 0}$ with parameters (P, Q), discriminant D, such that $\gcd(N, QD) = 1$, $(Q/N) = 1$, and $N|U_{\frac{1}{2}[N-(D/N)]}$. Then N is composite.

Similarly:

Let $N > 1$ be an odd integer. Assume that there exists a companion Lucas sequence $(V_n)_{n\geq 0}$, with parameters (P, Q), discriminant D, such that $N|QD$, $(Q/N) = -1$ and $N \nmid V_{\frac{1}{2}[N-(D/N)]}$. Then N is composite.

Proof. Indeed, if $N = p$ is an odd prime, not dividing QD and if $(Q/p) = 1$, then $p|U_{\psi(p)/2}$, and similarly, if $(Q/p) = -1$ then $p|V_{\psi(p)/2}$, as stated in (IV.23). In both cases there is a contradiction.
$\square$

And now I'm ready to present several tests; each one better than the preceding one.

Test 1. Let $N > 1$ be an odd integer and $N+1 = \prod_{i=1}^{s} q_i^{f_i}$. Assume that there exists an integer D such that $(D/N) = -1$, and for every prime factor q_i of $N+1$, there exists a Lucas sequence $(U_n^{(i)})_{n\geq 0}$ with discriminant $D = P_i^2 - 4Q_i$, where $\gcd(P_i, Q_i) = 1$, or $\gcd(N, Q_i) = 1$ and such that $N|U_{N+1}^{(i)}$ and $N \nmid U_{(N+1)/q_i}^{(i)}$. Then N is a prime.

Defect of this test: it requires the knowledge of all the prime factors of $N+1$ and the calculation of $U_n^{(i)}$ for $n = 1, 2, \ldots, N + 1$.

Proof. By (V.3), (V.4), $N|U_{\psi_D(N)}^{(i)}$ for every $i = 1, \ldots, s$. Let $\rho^{(i)}(N)$ be the smallest integer r such that $N|U_r^{(i)}$. By (IV.29) or (IV.22) and the hypothesis, $\rho^{(i)}(N)|(N+1)$, $\rho^{(i)}(N) \nmid (N+1)/q_i$, and also $\rho^{(i)}(N)|\psi_D(N)$. Hence $q_i^{f_i}|\rho^{(i)}(N)$ for every $i = 1, \ldots, s$. Therefore, $(N+1)|\psi_D(N)$ and by (V.2), N is a prime.
$\square$

The next test needs only half of the computations:

Test 2. Let $N > 1$ be an odd integer and $N+1 = \prod_{i=1}^{s} q_i^{f_i}$. Assume that there exists an integer D such that $(D/N) = -1$, and for every prime factor q_i of $N+1$, there exists a Lucas sequence $(V_n^{(i)})_{n\geq 0}$ with discriminant $D = P_i^2 - 4Q_i$, where $\gcd(P_i, Q_i) = 1$ or $\gcd(N, Q_i) = 1$, and such that $N|V_{(N+1)/2}^{(i)}$ and $N \nmid V_{(N+1)/2q_i}^{(i)}$. Then N is a prime.

Proof. By (IV.2), $N|U_{N+1}^{(i)}$. By (V.5), $N \nmid U_{(N+1)/q_i}^{(i)}$. By the test 1, N is a prime.
$\square$

The following tests will require only a partial factorization of $N+1$.

Test 3. Let $N > 1$ be an odd integer, let q be a prime factor of $N+1$ such that $2q > \sqrt{N} + 1$. Assume that there exists a Lucas sequence $(V_n)_{n\geq 0}$, with discriminant $D = P^2 - 4Q$, where $\gcd(P, Q) = 1$

or $\gcd(N, Q) = 1$, and such that $(D/N) = -1$, and $N|V_{(N+1)/2}$, $N \nmid V_{(N+1)/2q}$. Then N is a prime.

Defect of this test: it needs the knowledge of a fairly large prime factor of $N + 1$.

Proof. Let $N = \prod_{i=1}^{s} p_i^{e_i}$. By (IV.2), $N|U_{N+1}$, so by (IV.29) or (IV.22), $\rho(N)|(N+1)$. By (V.5), $N \nmid U_{(N+1)/q}$; hence, $\rho(N) \nmid (N+1)/q$, therefore $q|\rho(N)$. By (V.4) and (V.3), $N|U_{\psi_D(N)}$, so $\rho(N)$ divides $\psi_D(N)$, which in turn divides $N \prod_{i=1}^{s}(p_i - (D/p_i))$.

Since $q \nmid N$, then there exists p_i such that q divides $p_i - (D/p_i)$, thus $p_i \equiv (D/p_i) \pmod{2q}$. In conclusion, $p_i \geq 2q - 1 > \sqrt{N}$ and $1 \leq N/p_i < \sqrt{N} < 2q - 1$, and this implies that $N/p_i = 1$, that is N is a prime. $\qquad\square$

The next test, which was proposed by Morrison in 1975, may be viewed as the analogue of Pocklington's test indicated in Section III:

Test 4. Let $N > 1$ be an odd integer and $N + 1 = FR$, where $\gcd(F, R) = 1$ and the factorization of F is known. Assume that there exists D such that $(D/N) = -1$ and, for every prime q_i dividing F, there exists a Lucas sequence $(U_n^{(i)})_{n \geq 0}$ with discriminant $D = P_i^2 - 4Q_i$, where $\gcd(P_i, Q_i) = 1$, or $\gcd(N, Q_i) = 1$ and such that $N|U_{N+1}^{(i)}$ and $\gcd(U_{(N+1)/q_i}^{(i)}, N) = 1$. Then each prime factor p of N satisfies $p \equiv (D/p) \pmod{F}$. If, moreover, $F > \sqrt{N} + 1$, then N is a prime.

Proof. From the hypothesis, $\rho^{(i)}(N)|N+1$; a fortiori, $\rho^{(i)}(p)|N+1$. But $p \nmid U_{(N+1)/q}^{(i)}$, so $\rho^{(i)}(p)|(N+1)/q_i$, by (IV.29) or (IV.22). If $q_i^{f_i}$ is the exact power of q_i dividing F, then $q_i^{f_i}|\rho^{(i)}(p)$, so by (IV.18), $q_i^{f_i}$ divides $p - (D/p)$, and this implies that F divides $p - (D/p)$.

Finally, if $F > \sqrt{N} + 1$, then

$$p + 1 \geq p - (D/p) \geq F > \sqrt{N} + 1;$$

hence, $p > \sqrt{N}$. This implies that N itself is a prime. $\qquad\square$

The next result tells more about the possible prime factors of N.

Let N be an odd integer, $N + 1 = FR$, where $\gcd(F, R) = 1$ and the factorization of F is known. Assume that there exists a

Lucas sequence $(U_n)_{n\geq 0}$ with discriminant $D = P^2 - 4Q$, where $\gcd(P, Q) = 1$ or $\gcd(N, Q) = 1$ and such that $(D/N) = -1$, $N|U_{N+1}$, and $\gcd(U_F, N) = 1$. If p is a prime factor of N, then there exists a prime factor q of R such that $p \equiv (D/p) \pmod{q}$.

Proof. $\rho(p)|(p - (D/p))$ by (IV.18) and $\rho(p)|(N + 1)$. But $p \nmid U_F$, so $\rho(p) \nmid F$. Hence, $\gcd(\rho(p), R) \neq 1$ and there exists a prime q such that $q|R$ and $q|\rho(p)$; in particular, $p \equiv (D/p) \pmod{q}$.   □

This result is used in the following test:

Test 5. Let $N > 1$ be an odd integer and $N + 1 = FR$, where $\gcd(F, R) = 1$, the factorization of F is known, R has no prime factor less than B, where $BF > \sqrt{N} + 1$. Assume that there exists D such that $(D/N) = -1$ and the following conditions are satisfied:

(i) For every prime q_i dividing F, there exists a Lucas sequence $(U_n^{(i)})_{n\geq 0}$, with discriminant $D = P_i^2 - 4Q_i$, where $\gcd(P_i, Q_i) = 1$ or $\gcd(N, Q_i) = 1$ and such that $N|U_{N+1}^{(i)}$ and $\gcd(U_{(N+1)/q_i}^{(i)}, N) = 1$.

(ii) There exists a Lucas sequence $(U_n')_{n\geq 0}$, with discriminant $D = P'^2 - 4Q'$, where $\gcd(P', Q') = 1$ or $\gcd(N, Q') = 1$ and such that $N|U_{N+1}'$ and $\gcd(U_F', N) = 1$.

Then N is a prime.

Proof. Let p be a prime factor of N. By Test 4, $p \equiv (D/p) \pmod{F}$ and by the preceding result, there exists a prime factor q of R such that $p \equiv (D/p) \pmod{q}$. Hence, $p \equiv (D/p) \pmod{qF}$ and so,

$$p + 1 \geq p - (D/p) \geq qF \geq BF > \sqrt{N} + 1.$$

Therefore, $p > \sqrt{N}$ and N is a prime number.   □

The preceding test is more flexible than the others, since it requires only a partial factorization of $N + 1$ up to a point where it may be assured that the nonfactored part of $N + 1$ has no factors less than B.

Now I want to indicate, in a very succinct way, how to quickly calculate the terms of Lucas sequences with large indices. One of the methods is similar to that used in the calculations of high powers, which was indicated in Section III.

Write $n = n_0 2^k + n_1 2^{k-1} + \cdots + n_k$, with $n_i = 0$ or 1 and $n_0 = 1$; so $k = [(\log n)/(\log 2)]$. To calculate U_n (or V_n) it is necessary to perform the simultaneous calculation of U_m, V_m for various values of m. The following formulas are needed:

$$\begin{cases} U_{2j} = U_j V_j, \\ V_{2j} = V_j^2 - 2Q^j, \end{cases} \quad \text{[see formula (IV.2)]}$$

$$\begin{cases} 2U_{2j+1} = V_{2j} + PU_{2j}, \\ 2V_{2j+1} = PV_{2j} + DU_{2j}. \end{cases} \quad \text{[see formula (IV.5)]}$$

Put $s_0 = n_0 = 1$, and $s_{j+1} = 2s_j + n_{j+1}$. Then $s_k = n$. So, it suffices to calculate U_{s_j}, V_{s_j} for $j \leq k$; note that

$$U_{s_{j+1}} = U_{2s_j + n_{j+1}} = \begin{cases} U_{2s_j} & \text{or} \\ U_{2s_j+1}, \end{cases}$$

$$V_{s_{j+1}} = V_{2s_j + n_{j+1}} = \begin{cases} V_{2s_j} & \text{or} \\ V_{2s_j+1}. \end{cases}$$

Thus, it is sufficient to compute $2k$ numbers U_i and $2k$ numbers V_i, that is, only $4k$ numbers.

If it is needed to know U_n modulo N, then in all steps the numbers may be replaced by their least positive residues modulo N.

The second method is also very quick. For $j \geq 1$,

$$\begin{pmatrix} U_{j+1} & V_{j+1} \\ U_j & V_j \end{pmatrix} = \begin{pmatrix} P & -Q \\ 1 & 0 \end{pmatrix} \begin{pmatrix} U_j & V_j \\ U_{j-1} & V_{j-1} \end{pmatrix}.$$

If

$$M = \begin{pmatrix} P & -Q \\ 1 & 0 \end{pmatrix},$$

then

$$\begin{pmatrix} U_n & V_n \\ U_{n-1} & V_{n-1} \end{pmatrix} = M^{n-1} \begin{pmatrix} U_1 & V_1 \\ 0 & 2 \end{pmatrix}.$$

To find the powers of M, say M^m, write m in binary form and proceed in the manner followed to calculate a power of a number.

If U_n modulo N is to be determined, all the numbers appearing in the above calculation should be replaced by their least positive residues modulo N.

To conclude this section, I like to stress that there are many other primality tests of the same family, which are appropriate for numbers of certain forms, and use either Lucas sequences or other similar sequences.

Sometimes it is practical to combine tests involving Lucas sequences with the tests discussed in Section III; see the paper of Brillhart, Lehmer & Selfridge (1975). As a comment, I add (half-jokingly) the following rule of thumb: the longer the statement of the testing procedure, the quicker it leads to a decision about the primality.

The tests indicated so far are applicable to numbers of the form $2^n - 1$ (see Section VII on Mersenne numbers, where the test will be given explicitly), but also to numbers of the form $k \times 2^n - 1$ [see, for example, Inkeri's paper of 1960 or Riesel's book, (1985)].

VI. Fermat Numbers

For numbers having a special form, there are more suitable methods to test whether they are prime or composite.

The numbers of the form $2^m + 1$ were considered long ago.

If $2^m + 1$ is a prime, then m must be of the form $m = 2^n$, so it is a Fermat number $F_n = 2^{2^n} + 1$.

The Fermat numbers $F_0 = 3$, $F_1 = 5$, $F_2 = 17$, $F_3 = 257$, $F_4 = 65537$ are primes. Fermat believed and tried to prove that all Fermat numbers are primes. Since F_5 has 10 digits, in order to test its primality, it would be necessary to have a table of primes up to 100,000 (which was unavailable to him) or to derive and use some criterion for a number to be a factor of a Fermat number. This, Fermat failed to do.

Euler showed that every factor of F_n (with $n \geq 2$) must be of the form $k \times 2^{n+2} + 1$ and thus he discovered that 641 divides F_5:

$$F_5 = 641 \times 6700417.$$

Proof. It suffices to show that every prime factor p of F_n is of the form indicated. Since $2^{2^n} \equiv -1 \pmod{p}$, then $2^{2^{n+1}} \equiv 1 \pmod{p}$, so 2^{n+1} is the order of 2 modulo p. By Fermat's little theorem 2^{n+1}

divides $p - 1$; in particular, 8 divides $p - 1$. Therefore the Legendre symbol is $2^{(p-1)/2} \equiv (2/p) \equiv 1 \pmod{p}$, and so 2^{n+1} divides $(p - 1)/2$; this shows that $p = k \times 2^{n+2} + 1$. □

Since the numbers F_n increase very rapidly with n, it becomes laborious to check their primality.

Using the converse of Fermat's little theorem, as given by Lucas, Pepin obtained in 1877 a test for the primality of Fermat numbers. Namely:

Pepin's Test. Let $F_n = 2^{2^n} + 1$ (with $n \geq 2$) and $k \geq 2$. Then, the following conditions are equivalent:

(i) F_n is prime and $(k/F_n) = -1$.

(ii) $k^{(F_n-1)/2} \equiv -1 \pmod{F_n}$.

Proof. If (i) is assumed, then by Euler's criterion for the Legendre symbol

$$k^{(F_n-1)/2} \equiv \left(\frac{k}{F_n}\right) \equiv -1 \pmod{F_n}.$$

If, conversely, (ii) is supposed true, let a, $1 \leq a < F_n$, be such that $a \equiv k \pmod{F_n}$. Since $a^{(F_n-1)/2} \equiv -1 \pmod{F_n}$, then $a^{F_n-1} \equiv 1 \pmod{F_n}$. By Test 3 in Section III, F_n is prime. Hence

$$\left(\frac{k}{F_n}\right) \equiv k^{(F_n-1)/2} \equiv -1 \pmod{F_n}.$$

□

Possible choices of k are $k = 3, 5, 10$, because $F_n \equiv 2 \pmod 3$, $F_n \equiv 2 \pmod 5$, $F_n \equiv 1 \pmod 8$; hence, by Jacobi's reciprocity law

$$\left(\frac{3}{F_n}\right) = \left(\frac{F_n}{3}\right) = \left(\frac{2}{3}\right) = -1,$$

$$\left(\frac{5}{F_n}\right) = \left(\frac{F_n}{5}\right) = \left(\frac{2}{5}\right) = -1,$$

$$\left(\frac{10}{F_n}\right) = \left(\frac{2}{F_n}\right)\left(\frac{5}{F_n}\right) = -1.$$

This test is very practical in application. However, if F_n is composite, the test does not indicate any factor of F_n.

Lucas used it to show that F_6 is composite, and, in 1880, Landry showed that

$$F_6 = 274177 \times 67280421310721.$$

The factorization of F_7 was first performed by Morrison & Brillhart (1970, published in 1971), that of F_8 by Brent & Pollard (1981); see also Brent (1982).

RECORD

A. The largest known Fermat prime is $F_4 = 65537$.

B. The largest known composite Fermat number is F_{23471} (Keller, 1984, announced in 1985), which has the factor $5 \times 2^{23473} + 1$ and more than 10^{7000} digits. Keller has also shown in 1980 that F_{9448} is composite, having the factor $19 \times 2^{9450} + 1$.

C. Here are the composite Fermat numbers whose factorization is completely known: F_5, F_6, F_7, F_8, F_9, and F_{11}.

F_{11} was factored before F_9, because it has two small prime factors, which were known for a certain time. Brent found two more prime factors (by a method of factorization using elliptic curves) and he indicated that another factor of 564 digits was probably prime; this was confirmed by Morain (1988).

The factorization of F_9 was more difficult because it has no small prime factors; it was performed with the new method of number field sieve by the concerted efforts of Lenstra and Manasse (1990).

D. The smallest composite Fermat number for which no factor is known, is F_{14}. By means of Pepin's test, Selfridge & Hurwitz have shown in 1963 that F_{14} is composite, without finding any of its prime factors.

E. The smallest Fermat numbers for which it is not known if they are prime or composite, are F_{22}, F_{24}, F_{28}. In 1988, Young & Buell have published that F_{20} is composite; without indicating any factors (ten days of computation in a super computer, with a single bit answer!...).

The "most wanted" factorization is now that of F_{10}. It is already known that $F_{10} = 455925777 \times 6487031809 \times C291$, where Cm denotes a composite number with m digits.

Concerning these matters, a table of what is known appears in the book by Brillhart et al. (1983, and second edition in 1988). I wish also to recommend Riesel's book (1985), which is very well written and a source of information on matters of primality and factorization.

Here are some open problems:

(1) Are there infinitely many prime Fermat numbers?

This question became significant with the famous result of Gauss (see *Disquisitiones Arithmeticae* articles 365, 366—the last ones in the book—as a crowning result for much of the theory previously developed). He showed that if $n \geq 3$ is an integer, the regular polygon with n sides may be constructed by ruler and compass, if and only if $n = 2^k p_1 p_2 \cdots p_h$, where $k \geq 0$, $h \geq 0$ and $p_1, \ldots, p_h$ are distinct odd primes, each being a Fermat number.

In 1844, Eisenstein proposed, as a problem, to prove that there are indeed infinitely many prime Fermat numbers. I should add, that already in 1828, an anonymous writer stated that

$$2 + 1, \ 2^2 + 1, \ 2^{2^2} + 1, \ 2^{2^{2^2}} + 1, \ 2^{2^{2^{2^2}}} + 1, \ldots$$

are all primes, and in fact, they are the only prime Fermat numbers (apart from $2^{2^3} + 1$). However, as Selfridge showed, F_{16} is not a prime (being divisible by $3150 \times 2^{18} + 1$), and this fact disproved that conjecture.

(2) Are there infinitely many composite Fermat numbers?

Questions (1) and (2) seem beyond the reach of present-day methods and, side by side, they show how little is known on this matter.

(3) Is every Fermat number square-free (i.e. without square factors)?

It has been conjectured, for example by Lehmer and by Schinzel, that there exist infinitely many square-free Fermat numbers.

It is not difficult to show that if p is a prime number and p^2 divides some Fermat number, then $2^{p-1} \equiv 1 \pmod{p^2}$—this will be proved in detail in Chapter 5, Section III. Since Fermat numbers are pairwise relatively prime, if there exist infinitely many Fermat numbers with a square factor, then there exist infinitely many primes p satisfying the above congruence.

I shall discuss this congruence in Chapter 5. Let it be said here that it is very rarely satisfied. In particular, it is not known whether it holds infinitely often.

Sierpiński considered in 1958 the numbers of form $S_n = n^n + 1$, with $n \geq 2$.

He proved that if S_n is a prime, then there exists $m \geq 0$ such that $n = 2^{2^m}$, so S_n is a Fermat number $S_n = F_{m+2^m}$. It follows that the only numbers S_n which are primes and have less than 30,000,000,000 digits, are 5 and 257. Indeed, if $m = 0$, 1, one has $F_1 = 5$, $F_3 = 257$; if $m = 2$, one obtains F_6 and if $m = 3$, one gets F_{11}, both composite. For $m = 4$, one has F_{20}, which is now known to be composite. For $m = 5$, it is F_{37}, which is not known to be prime or composite. Since $2^{10} > 10^3$ then

$$F_{37} > 2^{2^{37}} > 2^{10^{11}} = (2^{10})^{10^{10}} > 10^{3 \times 10^{10}},$$

so F_{37} has more than 30,000,000,000 digits.

The primes of the form $n^n + 1$ are very rare. Are there only finitely many such primes? If so, there are infinitely many composite Fermat numbers. But all this is pure speculation, with no basis for any reasonable conjecture.

VII. Mersenne Numbers

If a number of the form $2^m - 1$ is a prime, then $m = q$ is a prime. Even more, it is not a difficult exercise to show that if $2^m - 1$ is a prime power, it must be a prime, and so m is a prime. [If you cannot do it alone, look at the paper of Ligh & Neal (1974).]

The numbers $M_q = 2^q - 1$ (with q prime) are called Mersenne numbers, and their consideration was motivated by the study of perfect numbers (see the addendum to this section).

Already at Mersenne's time, it was known that some Mersenne numbers are prime, others composite. For example, $M_2 = 3$, $M_3 = 7$, $M_5 = 31$, $M_7 = 127$ are primes, while $M_{11} = 23 \times 89$. In 1640, Mersenne stated that M_q is also a prime for $q = 13$, 17, 19, 31, 67, 127, 257; he was wrong about 67 and 257, and he did not include 61, 89, 107 (among those less than 257) which also produce Mersenne primes. Yet, his statement was quite astonishing, in view of the size of the numbers involved.

The obvious problem is to recognize if a Mersenne number is a prime, and if not, to determine its factors.

A classical result about factors was stated by Euler in 1750 and proved by Lagrange (1775) and again by Lucas (1878):

If q is a prime $q \equiv 3 \pmod 4$, then $2q + 1$ divides M_q if and only *if $2q + 1$ is a prime; in this case, if $q > 3$, then M_q is composite.*

Proof. Let $n = 2q + 1$ be a factor of M_q. Since $2^2 \not\equiv 1 \pmod{n}$, $2^q \not\equiv 1 \pmod{n}$, $2^{2q} - 1 = (2^q + 1)M_q \equiv 0 \pmod{n}$, then by Lucas test 2 (see Section III), n is a prime.

Conversely, let $p = 2q + 1$ be a prime. Since $p \equiv 7 \pmod 8$, then $(2/p) = 1$, so there exists m such that $2 \equiv m^2 \pmod p$. It follows that $2^q \equiv 2^{(p-1)/2} \equiv m^{p-1} \equiv 1 \pmod p$, so p divides M_q.

If, moreover, $q > 3$, then $M_q = 2^q - 1 > 2q + 1 = p$, so M_q is composite.   □

Thus if $q = 11, 23, 83, 131, 179, 191, 239, 251$, then M_q has the factor $23, 47, 167, 263, 359, 383, 479, 503$, respectively.

Around 1825, Sophie Germain considered, in connection with Fermat's last theorem, the primes q such that $2q + 1$ is also a prime. These primes are now called Sophie *Germain primes,* and I'll return to them in Chapter 5.

It is also very easy to determine the form of the factors of Mersenne numbers:

If n divides M_q, then $n \equiv \pm 1 \pmod 8$ and $n \equiv 1 \pmod q$.

Proof. It suffices to show that each prime factor p of M_q is of the form indicated.

If p divides $M_q = 2^q - 1$, then $2^q \equiv 1 \pmod q$; so by Fermat's little theorem, q divides $p - 1$, that is, $p - 1 = 2kq$ (since $p \neq 2$). So

$$\left(\frac{2}{p}\right) \equiv 2^{(p-1)/2} \equiv 2^{qk} \equiv 1 \pmod p,$$

therefore $p \equiv \pm 1 \pmod 8$, by the property of Legendre symbol already indicated in Section II.   □

The best method presently known to find out whether M_q is a prime or a composite number is based on the computation of a recurring sequence, indicated by Lucas (1878), and Lehmer (1930, 1935); see also Western (1932), Hardy & Wright (1938, p. 223), and Kaplansky (1945). However, explicit factors cannot be found in this manner.

If n is odd, $n \geq 3$, then $M_n = 2^n - 1 \equiv 7 \pmod{12}$. Also, if $N \equiv 7 \pmod{12}$, then the Jacobi symbol

$$\left(\frac{3}{N}\right) = \left(\frac{N}{3}\right)(-1)^{(N-1)/2} = -1.$$

Primality Test for Mersenne Numbers. Let $P = 2$, $Q = -2$, and consider the associated Lucas sequences $(U_m)_{m \geq 0}$, $(V_m)_{m \geq 0}$, which have discriminant $D = 12$. Then $N = M_n$ is a prime if and only if N divides $V_{(N+1)/2}$.

Proof. Let N be a prime. By (IV.2)

$$
\begin{aligned}
V_{(N+1)/2}^2 &= V_{N+1} + 2Q^{(N+1)/2} = V_{N+1} - 4(-2)^{(N-1)/2} \\
&\equiv V_{N+1} - 4\left(\frac{-2}{N}\right) \equiv V_{N+1} + 4 \pmod{N},
\end{aligned}
$$

because

$$\left(\frac{-2}{N}\right) = \left(\frac{-1}{N}\right)\left(\frac{2}{N}\right) = -1,$$

since $N \equiv 3 \pmod 4$ and $N \equiv 7 \pmod 8$. Thus it suffices to show that $V_{N+1} \equiv -4 \pmod{N}$.

By (IV.4), $2V_{N+1} = V_N V_1 + DU_N U_1 = 2V_N + 12U_N$; hence, by (IV.14) and (IV.13):

$$V_{N+1} = V_N + 6U_N \equiv 2 + 6(12/N) \equiv 2 - 6 \equiv -4 \pmod{N}.$$

Conversely, assume that N divides $V_{(N+1)/2}$. Then N divides U_{N+1} [by (IV.2)]. Also, by (IV.6) $V_{(N+1)/2}^2 - 12U_{(N+1)/2}^2 = 4(-1)^{(N+1)/2}$; hence, $\gcd(N, U_{(N+1)/2}) = 1$. Since $\gcd(N, 2) = 1$, then by the Test 1 (Section V)), N is a prime.   □

For the purpose of calculation, it is convenient to replace the Lucas sequence $(V_m)_{m \geq 0}$ by the following sequence $(S_k)_{k \geq 1}$, defined recursively as follows:

$$S_0 = 4, \qquad S_{k+1} = S_k^2 - 2;$$

thus the sequence begins with 4, 14, 194, ...
Then the test is phrased as follows:

$M_n = 2^n - 1$ *is prime if and only if* M_n *divides* S_{n-2}.

Proof. $S_0 = 4 = V_2/2$. Assume that $S_{k-1} = V_{2^k}/2^{2^{k-1}}$; then

$$S_k = S_{k-1}^2 - 2 = \frac{V_{2^k}^2}{2^{2^k}} - 2 = \frac{V_{2^{k+1}+2^{2^k}+1}}{2^{2^k}} - 2 = \frac{V_{2^{k+1}}}{2^{2^k}}.$$

By the test, M_n is prime if and only if M_n divides

$$V_{(M_n+1)/2} = V_{2^{n-1}} = 2^{2^{n-2}} S_{n-2},$$

or equivalently, M_n divides S_{n-2}.   □

The repetitive nature of the computations makes this test quite suitable. In this way, all examples of large Mersenne primes have been discovered. Lucas himself showed, in 1876, that M_{127} is a prime, while M_{67} is composite. Not much later, Pervushin showed that M_{61} is also a prime. Finally, in 1927 (published in 1932) Lehmer showed that M_{257} is also composite, settling one way or another, what Mersenne had asserted. Note that M_{127} has 39 digits and was the largest prime known before the age of computers.

RECORD

There are now 31 known Mersenne primes. Here is a complete list, with the discoverer and year of discovery (the recordman discoverer is Robinson):

q	Year	Discoverer
2	–	–
3	–	–
5	–	–
7	–	–
13	1461	Anonymous*
17	1588	P.A.Cataldi
19	1588	P.A.Cataldi
31	1750	L.Euler
61	1883	I.M.Pervushin
89	1911	R.E.Powers
107	1913	E. Fauquembergue
127	1876	E. Lucas
521	1952	R.M. Robinson
607	1952	R.M. Robinson
1279	1952	R.M. Robinson
2203	1952	R.M. Robinson
2281	1952	R.M. Robinson
3217	1957	H. Riesel
4253	1961	A. Hurwitz
4423	1961	A. Hurwitz
9689	1963	D.B. Gillies
9941	1963	D.B. Gillies
11213	1963	D.B. Gillies
19937	1971	B. Tuckerman
21701	1978	L.C. Noll & L. Nickel
23209	1979	L.C. Noll
44497	1979	H. Nelson & D. Slowinski
86243	1982	D. Slowinski
110503	1988	W.N. Colquitt & L. Welsch, Jr.
132049	1983	D. Slowinski
216091	1985	D. Slowinski

*See Dickson's *History of the Theory of Numbers*, Vol. I, p. 6.

The Mersenne primes with $q \leq 127$ were discovered before the computer age. Turing made, in 1951, the first attempt to find Mersenne primes using an electronic computer; however, he was unsuccessful.

In 1952, Robinson carried out Lucas's test using a computer SWAC (from the National Bureau of Standards in Los Angeles), with the assistance of D. H. and E. Lehmer. He discovered the Mersenne primes M_{521}, M_{607} on January 30, 1952—the first such discoveries with a computer. The primes M_{1279}, M_{2203}, M_{2281} were found later in the same year.

The largest known Mersenne prime is M_{216091}, it has 65050 digits and was discovered by Slowinski on September 1, 1985.

Until 1989, this was the largest known prime number, but it has now been dethroned (see Chapter 5, Section VI) by a non-Mersenne prime which must have made the poor Mersenne turn in his tomb.

The latest discovered Mersenne prime is M_{110503}; it was found by Colquitt & Welsch in 1988.

In 1989, Bateman, Selfridge & Wagstaff spelled out the following conjecture concerning Mersenne primes:

Let p be an odd natural number (not necessarily a prime, to begin with). If two of the following conditions are satisfied, so is the third one:

(a) $p = 2^k \pm 1$ or $p = 4^k \pm 3$

(b) M_p is a prime

(c) $\frac{2^p+1}{3}$ is a prime.

This conjecture has been checked to be true for all $p < 100000$.

The only primes $p < 100000$ for which the three conditions hold are: $p = 3, 5, 7, 13, 17, 19, 31, 61, 127$. It is conceivable that these are the only primes for which the three above conditions hold.

As already indicated, if q is a Sophie Germain prime and $q \equiv 3 \pmod 4$, then M_q is composite.

RECORD

The largest known Sophie Germain prime is $q = 39051 \times 2^{6001} - 1$ and M_q is the largest known composite Mersenne number (communicated by letter in 1987, by Keller). This provides the largest known composite Mersenne number M_q. I shall return to Sophie Germain primes in Chapter 5, Section II.

Riesel's book (1985) has a table of complete factorization of all numbers $M_n = 2^n - 1$, with n odd, $n \leq 257$. A more extensive table is in the book of Brillhart et al. (1983; see also the second edition, 1988).

Just as for Fermat numbers, there are many open problems about Mersenne numbers:

(1) Are there infinitely many Mersenne primes?

(2) Are there infinitely many composite Mersenne numbers?

The answer to both questions ought to be "yes", as I will try to justify. For example, I will indicate in Chapter 6, Section A, after (D5), that some sequences, similar to the sequence of Mersenne numbers, contain infinitely many composite numbers.

(3) Is every Mersenne number square-free?

Rotkiewicz showed in 1965 that if p is a prime and p^2 divides some Mersenne number, then $2^{p-1} \equiv 1 \pmod{p^2}$, the same congruence which already appeared in connection with Fermat numbers having a square factor.

I wish to mention two other problems involving Mersenne numbers, one of which has been solved, while the other one is still open.

Is it true that if M_q is a Mersenne prime, then M_{M_q} is also a prime number?

The answer is negative, since despite M_{13} being prime, $M_{M_{13}} = 2^{8191} - 1$ is composite [shown by Wheeler, see Robinson (1954)]. Note that $M_{M_{13}}$ has more than 2400 digits. In 1976, Keller discovered the prime factor

$$p = 2 \times 20644229 \times M_{13} + 1 = 338193759479$$

of the Mersenne number $M_{M_{13}}$, thus providing an easier verification that it is composite; only 13 squarings modulo p were needed to show that $2^{2^{13}} \equiv 2 \pmod{p}$. This has been kindly communicated to me by Keller in a recent letter, which contained also a wealth of other useful information. This factor is included in G. Haworth's *Mersenne Numbers*, quoted as reference 203 in Shanks' book *Solved and Unsolved Problems in Number Theory*, now in its third edition.

The second problem, by Catalan (1876), reported in Dickson's *History of the Theory Numbers*, Volume I, page 22, is the following.

Consider the sequence of numbers

$$
\begin{aligned}
C_1 &= 2^2 - 1 = 3 = M_2, \\
C_2 &= 2^{C_1} - 1 = 7 = M_3, \\
C_3 &= 2^{C_2} - 1 = 2^7 - 1 = 127 = M_7, \\
C_4 &= 2^{C_3} - 1 = 2^{127} - 1 = M_{127}, \ldots, C_{n+1} = 2^{C_n} - 1, \ldots .
\end{aligned}
$$

Are all numbers C_n primes? Are there infinitely many which are prime? At present, it is impossible to test C_5, which has more than 10^{38} digits!

ADDENDUM ON PERFECT NUMBERS

I shall consider now the perfect numbers, which are closely related to the Mersenne numbers.

A natural number $n > 1$ is said to be *perfect* if it is equal to the sum of all its aliquot parts, that is, its divisors d, with $d < n$. For example, $n = 6$, 28, 496, 8128 are the perfect numbers smaller than 10,000.

Perfect numbers were already known in ancient times. The first perfect number 6 was connected, by mystic and religious writers, to perfection, thus explaining that the Creation required 6 days, so PERFECT it is.

Euclid showed, in his *Elements*, Book IX, Proposition 36, that if q is a prime and $M_q = 2^q - 1$ is a prime, then $n = 2^{q-1}(2^q - 1)$ is a perfect number.

In a posthumous paper, Euler proved the converse: any even perfect number is of the form indicated by Euclid. Thus, the knowledge of even perfect numbers is equivalent to the knowledge of Mersenne primes.

And what about odd perfect numbers? Do they exist? Not even one has ever been found! This is a question which has been extensively searched, but its answer is still unknown.

Quick information on the progress made toward the solution of the problem may be found in Guy's book (1981) (quoted in General References), Section B1.

The methods to tackle the problem have been legion. I believe it is useful to describe them so the reader will get a feeling of what to do when nothing seems reasonable. The idea is to assume that there exists an odd perfect number N and to derive various consequences, concerning the number $\omega(N)$ of its distinct prime factors, the size of N, the multiplicative form, and the additive form of N, etc. I'll review what has been proved in each count.

(a) Number of distinct prime factors $\omega(N)$:

1980 (announced in 1975), Hagis: $\omega(N) \geq 8$. Chein proved also in his thesis, independently, that $\omega(N) \geq 8$ (this was in 1979).

1983 Hagis, and independently, Kishore: if $3 \nmid N$, then $\omega(N) \geq 11$.

Another result in this line was given by Dickson in 1913: for every $k \geq 1$ there are at most finitely many odd perfect numbers N, such that $\omega(N) = k$. In 1949, Shapiro gave a simpler proof.

Pomerance showed in 1977, for every $k \geq 1$: If the odd perfect number N has k distinct prime factors, then

$$N < (4k)^{(4k)^{2^{k^2}}}.$$

This gives an effective bound for such an odd perfect number (if any exists at all!)

(b) Lower bound for N:

Brent & Cohen (1989) have established that $N > 10^{160}$. In a forthcoming paper, which includes also teRiele as a co-author, the bound is $N > 10^{300}$. (The statement of Buxton & Elmore of 1976, that $N > 10^{200}$ has not been substantiated in detail, and so it is not acceptable).

(c) Multiplicative structure of N:

The first result is by Euler. $N = p^e k^2$, where p is a prime not dividing k, and $p \equiv e \equiv 1 \pmod{4}$.

There have been numerous results on the kind of number k.

For example, in 1972 Hagis & McDaniel showed that k is not a cube.

(d) Largest prime factor of N:

In 1975, Hagis & McDaniel showed that the largest prime factor of N should be greater than 100110.

Already in 1955, Muskat had even shown better: there must be a prime factor larger than 10^8. However, his result, which was based on calculations of Mills, remained unpublished.

(e) Largest prime-power factor of N:

In 1966, Muskat showed that N must have a prime-power factor greater than 10^{12}; this has been improved by Cohen (1987) to 10^{20}.

(f) Other prime factors of N:

In 1975, Pomerance showed that the second largest prime factor of N should be at least 139.

In 1952, Grün showed that the smallest prime factor p_1 of N should satisfy the relation $p_1 < \frac{2}{3}\omega(N) + 2$.

(g) Ore's conjecture:

In 1948, Ore considered the harmonic mean of the divisors of N, namely,

$$H(N) = \frac{\tau(N)}{\sum_{d|N} \frac{1}{d}}$$

where $\tau(N)$ denotes the number of divisors of N.

If N is a perfect number, then $H(N)$ is an integer; indeed, whether N is even or odd, this follows from Euler's results.

Actually, Laborde noted in 1955, that N is an even perfect number if and only if $N = 2^{H(N)-1}(2^{H(N)} - 1)$, hence $H(N)$ is an integer, and in fact a prime.

Ore conjectured that if N is odd, then $H(N)$ is not an integer. The truth of this conjecture would imply, therefore, that there do not exist odd perfect numbers.

Ore verified that the conjecture is true if N is a prime-power or if $N < 10^4$. Since 1954 (published only in 1972), Mills checked its truth for $N < 10^7$, as well as for numbers of special form, in particular, if all prime-power factors of N are smaller than 65551^2.

Pomerance (unpublished) verified Ore's conjecture when $\omega(N) \le 2$, by showing that if $\omega(N) \le 2$ and $H(N)$ is an integer, then N is an even perfect number (kindly communicated to me by letter).

The next results do not distinguish between even or odd perfect numbers. They concern the distribution of perfect numbers. The idea is to define, for every $x \ge 1$, the function $V(x)$, which counts the perfect numbers less or equal to x:

$$V(x) = \#\{\, N \text{ perfect} \mid N \le x \,\}.$$

The limit $\lim_{x \to \infty} V(x)/x$ represents a natural density for the set of perfect numbers. In 1954, Kanold showed the $\lim_{x \to \infty} V(x)/x = 0$. Thus, $V(x)$ grows to infinity slower than x does.

The following more precise result of Wirsing (1959) tells how slow $V(x)$ grows: there exist x_0 and $C > 0$ such that if $x \geq x_0$ then

$$V(x) \leq e^{\frac{C \log x}{\log \log x}}.$$

In particular, for every $\varepsilon > 0$ there exists a positive constant C such that $V(x) < Cx^{\varepsilon}$.

All the results that I have indicated about the problem of the existence of odd perfect numbers represent a considerable amount of work, sometimes difficult and delicate. Yet, I believe the problem stands like an unconquerable fortress. For all that is known, it would be almost by luck that an odd perfect number would be found. On the other hand, nothing that has been proved is decisive to show that odd perfect numbers do not exist. New ideas are required.

I wish to conclude this overview of perfect numbers with the following results of Sinha (1974) (the proof is elementary and should be an amusing exercise—just get your pencil ready!):

28 is the only even perfect number that is of the form $a^n + b^n$ with $n \geq 2$, and $\gcd(a, b) = 1$. It is also the only even perfect number of the form $a^n + 1$, with $n \geq 2$. And finally, there is no even perfect number of the form

$$a^{n^{n^{\cdot^{\cdot^{\cdot^{n}}}}}} + 1$$

with $n \geq 2$ and at least two exponents n.

VIII. Pseudoprimes

In this section I shall consider composite numbers having a property which one would think that only prime numbers possess.

A. PSEUDOPRIMES IN BASE 2 (psp)

According to Fermat's little theorem, every odd prime number p satisfies the congruence $2^{p-1} \equiv 1 \pmod{p}$.

However, in 1819 Sarrus observed that $341 = 11 \times 31$ satisfies the congruence $2^{340} \equiv 1 \pmod{341}$. This means that the converse of Fermat's little theorem is not true.

A *pseudoprime* (in base 2), also called a *Poulet number*, is a composite odd number n such that $2^{n-1} \equiv 1 \pmod{n}$.

I abbreviate: psp or psp(2). Already in 1926, Poulet determined all the pseudoprimes less than 5×10^7; he extended his table in 1938, up to 10^8. Thus, the smallest pseudoprimes are 341, 561, 645, 1105, 1387, 1729, 1905.

It is clear that if n is such that $2^{n-1} \not\equiv 1 \pmod{n}$, then n is not a prime (nor a pseudoprime). This remark is useful in trying to investigate if a given number is prime and I shall discuss it more amply in the sequel.

In order to know more about primes, it is natural to study the pseudoprimes.

Suppose I would like to write a chapter about pseudoprimes for the *Guinness Book of Records*. How would I organize it?

The natural questions should be basically the same as those for prime numbers. For example: How many pseudoprimes are there? Can one tell whether a number is a pseudoprime? Are there ways of generating pseudoprimes? How are the pseudoprimes distributed?

As it turns out, not surprisingly, there are infinitely many pseudoprimes, and there are many ways to generate infinite sequences of pseudoprimes.

The simplest proof was given in 1903 by Malo, who showed that if n is a pseudoprime, and if $n' = 2^n - 1$, then n' is also a pseudoprime. Indeed, n' is obviously composite, because if $n = ab$ with $1 < a$, $b < n$, then $2^n - 1 = (2^a - 1)(2^{a(b-1)} + 2^{a(b-2)} + \cdots + 2^a + 1)$. Also n divides $2^{n-1} - 1$, hence n divides $2^n - 2 = n' - 1$; so $n' = 2^n - 1$ divides $2^{n'-1} - 1$.

In 1904, Cipolla gave another proof, using the Fermat numbers:

If $m > n > \cdots > s > 1$ are integers and N is the product of the Fermat numbers $N = F_m F_n \cdots F_s$, then N is a pseudoprime if and only if $2^s > m$.

Indeed, the order of 2 modulo N is 2^{m+1}, which is equal to the least common multiple of the orders $2^{m+1}, 2^{n+1}, \ldots, 2^{s+1}$ of 2 modulo each factor factor $F_m, F_n, \ldots, F_s$ of N. Thus $2^{N-1} \equiv 1 \pmod{N}$ if and only if $N-1$ is divisible by 2^{m+1}. But $N-1 = F_m F_n \cdots F_s - 1 = 2^{2^s} Q$, where Q is an odd integer. Thus, the required condition is $2^s > m$. $\square$

As it was indicated in Chapter 1, the Fermat numbers are pairwise relatively prime, so the above method leads to pairwise relatively prime pseudoprimes. See below, for another method also indicated by Cipolla.

There also exist even composite integers satisfying the congruence $2^n \equiv 2 \pmod{n}$—they may be called *even pseudoprimes*. The smallest one is $m = 2 \times 73 \times 1103 = 161038$, discovered by Lehmer in 1950. In 1951, Beeger showed the existence of infinitely many even pseudoprimes; each one must have at least two odd prime factors.

How "far" are pseudoprimes from being primes? From Cipolla's result, there are pseudoprimes with arbitrarily many prime factors. This is not an accident. In fact, in 1949, Erdös proved that for every $k \geq 2$ there exist infinitely many pseudoprimes, which are the product of exactly k distinct primes.

Here is an open question: Are there infinitely many integers $n > 1$ such that

$$2^{n-1} \equiv 1 \pmod{n^2}?$$

This is equivalent to each of the following problems (see Rotkiewicz, 1965):

Are there infinitely many pseudoprimes that are squares?

Are there infinitely many primes p such that $2^{p-1} \equiv 1 \pmod{p^2}$?

This congruence was already encountered in the question of square factors of Fermat numbers and Mersenne numbers. I shall return to primes of this kind in Chapter 5, Section III.

On the other hand, a pseudoprime need not be square-free. The smallest such examples are $1{,}194{,}649 = 1093^2$, $12{,}327{,}121 = 3511^2$, $3{,}914{,}864{,}773 = 29 \times 113 \times 1093^2$.

B. PSEUDOPRIMES IN BASE $a(\text{psp}(a))$

It is also useful to consider the congruence $a^{n-1} \equiv 1 \pmod{n}$, for $a \neq 1$. If n is a prime and $1 < a < n$, then the above congruence holds necessarily. So, if, for example, $2^{n-1} \equiv 1 \pmod{n}$, but, say, $3^{n-1} \not\equiv 1 \pmod{n}$, then n is not a prime.

This leads to the more general study of the *a-pseudoprimes* n, namely, the composite integers $n > a$ such that $a^{n-1} \equiv 1 \pmod{n}$.

In 1904, Cipolla indicated also how to obtain a-pseudoprimes. Let $a \geq 2$, let p be any odd prime such that p does not divide $a(a^2 - 1)$. Let

$$n_1 = \frac{a^p - 1}{a - 1}, \quad n_2 = \frac{a^p + 1}{a + 1}, \quad n = n_1 n_2;$$

then n_1 and n_2 are odd and n is composite. Since $n_1 \equiv 1 \pmod{2p}$ and $n_2 \equiv 1 \pmod{2p}$, then $n \equiv 1 \pmod{2p}$. From $a^{2p} \equiv 1 \pmod{n}$ it follows that $a^{n-1} \equiv 1 \pmod{n}$, so n is an a-pseudoprime.

Since there exist infinitely many primes, then there also exist infinitely many a-pseudoprimes (also when $a > 2$).

There are other methods in the literature to produce very quickly increasing sequences of a-pseudoprimes.

Crocker showed in 1962 how to generate other infinite sequences of a-pseudoprimes. Let a be even, but not of the form 2^{2^r}, with $r \geq 0$. Then, for every $n \geq 1$, the number $a^{a^n} + 1$ is an a-pseudoprime.

It is therefore futile to wish to write the largest a-pseudoprime.

In 1959, Schinzel showed that for every $a > 1$, there exist infinitely many pseudoprimes in base a that are products of two distinct primes.

In 1971, in his thesis, Lieuwens extended simultaneously this result of Schinzel and Erdös' result about pseudoprimes in base 2: for every $k \geq 2$ and $a > 1$, there exist infinitely many pseudoprimes in base a, which are products of exactly k distinct primes.

In 1972, Rotkiewicz showed that if $p \geq 2$ is a prime not dividing $a \geq 2$, then there exist infinitely many pseudoprimes in base a that are multiples of p; the special case when $p = 2$ dates back to 1959, also by Rotkiewicz.

Here is a table, from the paper of Pomerance, Selfridge & Wagstaff (1980), which gives the smallest pseudoprimes for various bases, or simultaneous bases.

Bases	Smallest psp		
2	341	=	11×31
3	91	=	7×13
5	217	=	7×31
7	25	=	5×5
2,3	1105	=	$5 \times 13 \times 17$
2,5	561	=	$3 \times 11 \times 17$
2,7	561	=	$3 \times 11 \times 17$
3,5	1541	=	23×67
3,7	703	=	19×37
5,7	561	=	$3 \times 11 \times 17$
2,3,5	1729	=	$7 \times 13 \times 19$
2,3,7	1105	=	$5 \times 13 \times 17$
2,5,7	561	=	$3 \times 11 \times 17$
3,5,7	29341	=	$13 \times 37 \times 61$
2,3,5,7	29341	=	$13 \times 37 \times 61$

To be a pseudoprime for different bases, like 561 (for the bases 2,

5, 7) is not abnormal. Indeed, Baillie & Wagstaff and Monier showed independently (1980):

If n is a composite number, then the number $B_{\text{psp}}(n)$ of bases a, $1 < a \leq n - 1$, for which n is a psp(a), is equal to $\{\prod_{p|n} \gcd(n - 1, p - 1)\} - 1$.

It follows that if n is an odd composite number, which is not a power of 3, then n is a pseudoprime for at least two bases a, $1 < a \leq n - 1$.

It will be seen in Section IX that there exist composite numbers n, which are pseudoprimes for all bases a, $1 < a < n$.

By experimentation, it is observed that, usually, $B_{\text{psp}}(n)$ is very small compared to n, and, in fact, a statement of this kind was shown by Pomerance (it is in Baillie & Wagstaff's paper).

As I have said, if there exists a such that $1 < a < n$ and $a^{n-1} \not\equiv 1$ (mod n), then n is composite, but not conversely. This gives therefore a very practical way to ascertain that many numbers are composite. There are other congruence properties, similar to the above, which give also easy methods to discover that certain numbers are composite.

I shall describe several of these properties; their study has been justified by the problem of primality testing. As a matter of fact, without saying it explicitly, I have already considered these properties in Sections III and V. First, there are properties about the congruence $a^m \equiv 1$ (mod n), which lead to the Euler a-pseudoprimes and strong a-pseudoprimes. In another section, I will examine the Lucas pseudoprimes, which concern congruence properties satisfied by terms of Lucas sequences.

C. EULER PSEUDOPRIMES IN BASE a (epsp(a))

According to Euler's congruence for the Legendre symbol, if $a \geq 2$, p is a prime and p does not divide a, then

$$\left(\frac{a}{p}\right) \equiv a^{(p-1)/2} \pmod{p}.$$

This leads to the notion of an *Euler pseudoprime in base a* (epsp(a)), proposed by Shanks in 1962. These are odd composite numbers n, such that $\gcd(a, n) = 1$ and the Jacobi symbol satisfies the congruence

$$\left(\frac{a}{n}\right) \equiv a^{(n-1)/2} \pmod{n}.$$

Clearly, every epsp(a) is an a-pseudoprime.

There are many natural questions about epsp(a) which I enumerate now:

(e1) Are there infinitely many epsp(a), for each a?

(e2) Are there epsp(a) with arbitrary large number of distinct prime factors, for each a?

(e3) For every $k \geq 2$ and base a, are there infinitely many epsp(a), which are equal to the product of exactly k distinct prime factors?

(e4) Can an odd composite number be an epsp(a) for every possible a, $1 < a < n$, $\gcd(a, n) = 1$?

(e5) For how many bases a, $1 < a < n$, $\gcd(a, n) = 1$, can the number n be an epsp(a)?

In 1986, Kiss, Phong & Lieuwens showed that given $a \geq 2$, $k \geq 2$, and $d \geq 2$, there exist infinitely many epsp(a), which are the product of k distinct primes and are congruent to 1 modulo d.

This gives a strong affirmative answer to (e3), and therefore also to (e2) and (e1).

In 1976, Lehmer showed that if n is odd composite, then it cannot be an epsp(a), for every a, $1 < a < n$, $\gcd(a, n) = 1$. So the answer to (e4) is negative.

In fact, more is true, as shown by Solovay & Strassen in 1977: a composite integer n can be an Euler pseudoprime for at most $\frac{1}{2}\phi(n)$ bases a, $1 < a < n$, $\gcd(a, n) = 1$. This gives an answer to question (e5). The proof is immediate, noting that the residue classes $a \bmod n$, for which $(a/n) \equiv a^{(n-1)/2}$ (mod n) form a subgroup of $(Z/n)^{\times}$ (group of invertible residue classes modulo n), which is a proper subgroup (by Lehmer's result); hence it has at most $\frac{1}{2}\phi(n)$ elements—by dear old Lagrange's theorem.

Let n be an odd composite integer. Denote by $B_{\mathrm{epsp}}(n)$ the number of bases a, $1 < a < n$, $\gcd(a, n) = 1$, such that n is epsp(a). Monier showed in 1980 that

$$B_{\mathrm{epsp}}(n) = \delta(n) \prod_{p|n} \gcd\left(\frac{n-1}{2}, p-1\right) - 1.$$

Here

$$
\delta(n) = \begin{cases}
2 & \text{if } v_2(n) - 1 = \min_{p|n}\{\, v_2(p-1)\,\}, \\
\frac{1}{2} & \text{if there exists a prime } p \text{ dividing } n \text{ such that } v_{p(n)} \text{ is odd} \\
& \text{and } v_2(p-1) < v_2(n-1), \\
1 & \text{otherwise,}
\end{cases}
$$

and for any integer m and prime p, $v_p(m)$ denotes the exponent of p in the factorization of m, that is, the p-adic value of m.

D. STRONG PSEUDOPRIMES IN BASE a (spsp(a))

A related property is the following: Let n be an odd composite integer, let $n - 1 = 2^s d$, with d odd and $s \geq 1$; let a be such that $1 < a < n$, $\gcd(a, n) = 1$.

n is a *strong pseudoprime in base a* (spsp(a)) if $a^d \equiv 1 \pmod{n}$ or $a^{2^r d} \equiv -1 \pmod{n}$, for some r, $0 \leq r < s$.

Note that if n is a prime, then it satisfies the above condition, for every a, $1 < a < n$, $\gcd(a, n) = 1$.

Selfridge showed (see the proof in Williams' paper, 1978) that every spsp(a) is an epsp(a).

There are partial converses.

By Malm (1977): if $n \equiv 3 \pmod 4$ and n is an epsp(a), then n is a spsp(a).

By Pomerance, Selfridge & Wagstaff (1980): if n is odd, $(a/n) = -1$ and n is an epsp(a), then n is also a spsp(a). In particular, if $n \equiv 5 \pmod 8$ and n is an epsp(2), then it is a spsp(2).

Concerning the strong pseudoprimes, I may ask questions (s1)–(s5), analogous to the questions about Euler pseudoprimes given in Section VIII C.

In 1980, Pomerance, Selfridge & Wagstaff proved that for every base $a > 1$, there exist infinitely many spsp(a), and this answers in the affirmative, questions (s1), as well as (e1). I shall say more about this in the study of the distribution of pseudoprimes (Chapter 4, Section VI).

For base 2, it is possible to give infinitely many spsp(2) explicitly, as I indicate now.

If n is a psp(2), then $2^n - 1$ is a spsp(2). Since there are infinitely many psp(2), this gives explicitly infinitely many spsp(2); among these are all composite Mersenne numbers. It is also easy to see that if a Fermat number is composite, then it is a spsp(2).

Similarly, since there exist pseudoprimes with arbitrarily large numbers of distinct prime factors, then (s2), as well as (e2), have a positive answer; just note that if $p_1, p_2, \ldots, p_k$ divide the pseudoprime n, then $2^{p_i} - 1$ $(i = 1, \ldots, k)$ divides the spsp(2)$2^n - 1$.

In virtue of Lehmer's negative answer to (e4) and Selfridge's result, then clearly (s4) has also a negative answer. Very important —as I shall indicate later, in connection with the Monte Carlo primality testing methods—is the next theorem by Rabin, corresponding to Solovay & Strassen's result for Euler pseudoprimes. And it is tricky to prove:

If $n > 4$ is composite, there are at least $3(n - 1)/4$ integers a, $1 < a < n$, for which n is not a spsp(a). So, the number of bases a, $1 < a < n$, $\gcd(a, n) = 1$, for which an odd composite integer is spsp(a), is at most $(n - 1)/4$. This answers question (s5).

Monier (1980) has also determined a formula for the number $B_{\mathrm{spsp}}(n)$, of bases a, $1 < a < n$, $\gcd(a, n) = 1$, for which the odd composite integer n is spsp(a). Namely:

$$B_{\mathrm{spsp}}(n) = \left[1 + \frac{2^{\omega(n)\nu(n)} - 1}{2^{\omega(n)} - 1}\right] \left[\prod_{p|n} \gcd(n^*, p^*)\right] - 1,$$

where $\omega(n)$ = number of distinct prime factors of n,

$\quad v_p(m)$ = exponent of p in the factorization of m (any natural number),

$\quad m^*$ = largest odd divisor of $m - 1$,

$$\nu(n) = \min_{p|n}\{\, v_2(p - 1)\,\}.$$

Just for the record, the smallest spsp(2) is $2047 = 23 \times 89$.

The smallest number which is a strong pseudoprime relative to the bases 2 and 3 is $1{,}373{,}653 = 829 \times 1657$; and relative to the bases 2, 3, and 5 is $25{,}326{,}001 = 2251 \times 11251$. There are only 13 such numbers less than 25×10^9.

The following table is taken from the paper by the 3 Knights of Numerology:

Numbers Less Than 25×10^9, which are spsp in Bases 2, 3, 5

| Number | psp to base | | | Factorization |
	7	11	13	
25,326,001	no	no	no	2251×11251
161,304,001	no	spsp	no	7333×21997
960,946,321	no	no	no	11717×82013
1,157,839,381	no	no	no	24061×48121
3,215,031,751	spsp	psp	psp	$151 \times 751 \times 28351$
3,697,278,427	no	no	no	30403×121609
5,764,643,587	no	no	spsp	37963×151849
6,770,862,367	no	no	no	41143×164569
14,386,156,093	psp	psp	psp	$397 \times 4357 \times 8317$
15,579,919,981	psp	spsp	no	88261×176521
18,459,366,157	no	no	no	67933×271729
19,887,974,881	psp	no	no	81421×244261
21,276,028,621	no	psp	psp	103141×206281

To this table, I add the list of pseudoprimes up to 25×10^9 which are not square-free and their factorizations:

$$1194649 = 1093^2,$$
$$12327121 = 3511^2,$$
$$3914864773 = 29 \times 113 \times 1093^2,$$
$$5654273717 = 1093^2 \times 4733,$$
$$6523978189 = 43 \times 127 \times 1093^2,$$
$$22178658685 = 5 \times 47 \times 79 \times 1093^2.$$

With the exception of the last two, the numbers in the above list are strong pseudoprimes.

Note that the only prime factors to the square are 1093, 3511. The occurrence of these numbers will be explained in Chapter 5, Section III.

IX. Carmichael Numbers

In 1912, Carmichael considered a more rare kind of numbers. They are the composite numbers n such that $a^{n-1} \equiv 1 \pmod{n}$ for every integer a, $1 < a < n$, such that a is relatively prime to n.

The smallest Carmichael number is $561 = 3 \times 11 \times 17$.

I shall now indicate a characterization of Carmichael numbers.

Recall that I have introduced, in Section II, Carmichael's function $\lambda(n)$, which is the maximum of the orders of $a \bmod n$, for every a, $1 \le a < n$, $\gcd(a, n) = 1$; in particular, $\lambda(n)$ divides $\phi(n)$.

Carmichael showed that n is a Carmichael number if and only if n is composite and $\lambda(n)$ divides $n - 1$. (It is the same as saying that if p is any prime dividing n then $p - 1$ divides $n - 1$.)

It follows that every Carmichael number is odd and the product of three or more distinct prime numbers.

Explicitly, if $n = p_1 p_2 \cdots p_r$ (product of distinct primes), then n is a Carmichael number if and only if $p_i - 1$ divides $n/p_i - 1$ (for $i = 1, 2, \ldots, r$).

Therefore, if n is a Carmichael number, then also $a^n \equiv a \pmod{n}$, for every integer $a \ge 1$.

Schinzel noted in 1959 that, for every $a \ge 2$ the smallest pseudoprime m_a in base a, satisfies necessarily $m_a \le 561$. Moreover, there exists a such that $m_a = 561$. Explicitly, let p_i $(i = 1, \ldots, s)$ be the primes such that $2 < p_{e_i} < 561$; for each p_i let e_i be such that $p_{e_i}^i < 561 < p_{e_i+1}^i$; let g_i be a primitive root modulo $p_{e_i}^i$, and by the Chinese remainder theorem, let a be such that $a \equiv 3 \pmod{4}$ and $a \equiv g_i \pmod{p_{e_i}^i}$ for $i = 1, \ldots, s$. Then $m_a = 561$.

Carmichael and Lehmer determined the smallest Carmichael numbers:

$561 = 3 \times 11 \times 17$	$15841 = 7 \times 31 \times 73$	$101101 = 7 \times 11 \times 13 \times 101$
$1105 = 5 \times 13 \times 17$	$29341 = 13 \times 37 \times 61$	$115921 = 13 \times 37 \times 241$
$1729 = 7 \times 13 \times 19$	$41041 = 7 \times 11 \times 13 \times 41$	$126217 = 7 \times 13 \times 19 \times 73$
$2465 = 5 \times 17 \times 29$	$46657 = 13 \times 37 \times 97$	$162401 = 17 \times 41 \times 233$
$2821 = 7 \times 13 \times 31$	$52633 = 7 \times 73 \times 103$	$172081 = 7 \times 13 \times 31 \times 61$
$6601 = 7 \times 23 \times 41$	$62745 = 3 \times 5 \times 47 \times 89$	$188461 = 7 \times 13 \times 19 \times 109$
$8911 = 7 \times 19 \times 67$	$63973 = 7 \times 13 \times 19 \times 37$	$252601 = 41 \times 61 \times 101$
$10585 = 5 \times 29 \times 73$	$75361 = 11 \times 17 \times 31$	

Open problem: Are there infinitely many Carmichael numbers?

As I have mentioned before, even the more restricted problem (Duparc, 1952) is open: Are there infinitely many composite integers n such that $2^{n-1} \equiv 1 \pmod{n}$ and $3^{n-1} \equiv 1 \pmod{n}$?

I have also considered in Section II the question of existence of composite n such that $\phi(n)$ divides $n - 1$. Since $\lambda(n)$ divides $\phi(n)$, any such n must be a Carmichael number.

As it is not known whether there exist infinitely many Carmichael numbers, a fortiori, it is not known whether there exist Carmichael numbers with arbitrarily large number of prime factors. In this respect, there is a result of Duparc (1952) (see also Beeger, 1950):

For every $r \geq 3$, there exists only finitely many Carmichael numbers with r prime factors, of which the smallest $r - 2$ factors are given in advance. I shall return to these questions in Chapter 4.

In 1939, Chernick gave the following method to obtain Carmichael numbers. Let $m \geq 1$ and $M_3(m) = (6m + 1)(12m + 1)(18m + 1)$. If m is such that all three factors above are prime, then $M_3(m)$ is a Carmichael number. This yields Carmichael numbers with three prime factors.

Similarly, if $k \geq 4$, $m \geq 1$, let

$$M_k(m) = (6m + 1)(12m + 1) \prod_{i=1}^{k-2} (9 \times 2^i m + 1).$$

If m is such that all k factors are prime numbers and, moreover, 2^{k-4} divides m, then $M_k(m)$ is a Carmichael number with k prime factors. This method has been used by Wagstaff and Yorinaga to obtain large Carmichael numbers or Carmichael numbers with many factors.

RECORD

The largest known Carmichael numbers have been recently found by Dubner (1989), with a modification of the method of Chernick. These numbers are products pqr of three primes, with $p = 6m + 1$, $q = 12m + 1$, $r = \frac{pq-1}{x} + 1$. The largest known Carmichael number has 3710 digits and it was obtained by the above method with the parameters

$$t = 3 \times 5 \times 7 \times 11 \times \cdots \times 43 \times 47 = 307,444,891,294,245,705$$

$$c = 141847, \quad m = \frac{(tc-1)^{41}}{4}, \quad x = 123165.$$

Then p, q, r are primes with respectively 929, 929 and 1853 digits and pqr is a Carmichael number with 3710 digits. The computation took twenty hours. The same computations yielded several other Carmichael numbers of comparable size.

The previous records were Carmichael numbers of the type $M_3(m)$ and had been found by

Dubner in 1985 (1057 digits),

Woods & Huenemann in 1982 (432 digits),

Atkin in 1980 (370 digits),

Wagstaff in 1980 (321 digits).

Concerning Carmichael numbers with more than 3 factors, here are the largest ones which I know, found by Dubner: $M_4(m)$, with $m = \frac{1}{6}323323 \times 655899 \times 10^{40}$, with 270 digits;

$M_5(m)$, with $m = \frac{1}{6}323323 \times 426133 \times 10^{16}$, with 139 digits;

$M_6(m)$, with $m = \frac{1}{6}323323 \times 239556 \times 10^7$, with 112 digits;

$M_7(m)$ with $m = 323323 \times 160 \times 8033$, with 93 digits.

In 1978, Yorinaga found eight Carmichael numbers with 13 prime factors—these are the ones with the largest number of prime factors known up to now.

I shall discuss the distribution of Carmichael numbers in Chapter 4, Section VIII.

X. Lucas Pseudoprimes

In view of the analogy between sequences of binomials $a^n - 1$ ($n \geq 1$) and Lucas sequences, it is no surprise that pseudoprimes should have a counterpart involving Lucas sequences. For each parameter $a \geq 2$, there were the a-pseudoprimes and their cohort of Euler pseudoprimes and strong pseudoprimes in base a. In this section, to all pairs (P, Q) of nonzero integers, will be associated the corresponding Lucas pseudoprimes, the Euler–Lucas pseudoprimes, and the strong Lucas pseudoprimes. Their use will parallel that of pseudoprimes.

Let P, Q be nonzero integers, $D = P^2 - 4Q$ and consider the associated Lucas sequences $(U_n)_{n \geq 0}$, $(V_n)_{n \geq 0}$.

Recall (from Section IV) that if n is an odd prime, then:

(X.1) If $\gcd(n, D) = 1$, then $U_{n-(D/n)} \equiv 0 \pmod{n}$.

(X.2) $U_n \equiv (D/n) \pmod{n}$.

(X.3) $V_n \equiv P \pmod{n}$.

(X.4) If $\gcd(n, D) = 1$, then $V_{n-(D/n)} \equiv 2Q^{(1-(D/n))/2} \pmod{n}$.

If n is an odd composite number and the congruence (X.1) holds, then n is called a *Lucas pseudoprime (with the parameters (P, Q))*, abbreviated $\ell\,\mathrm{psp}(P, Q)$.

It is alright to make such a definition, but do these numbers exist?
If so, are they worthwhile to study?

A. FIBONACCI PSEUDOPRIMES

To begin, it is interesting to look at the special case of Fibonacci
numbers, where $P = 1$, $Q = -1$, $D = 5$. In this situation, it is more
appropriate to call *Fibonacci pseudoprimes* the $\ell \, \text{psp}(1, -1)$.

The smallest Fibonacci pseudoprimes are $323 = 17 \times 19$ and $377 =
13 \times 29$; indeed, $(5/323) = (5/377) = -1$ and it may be calculated
that $U_{324} \equiv 0 \pmod{323}$, $U_{378} \equiv 0 \pmod{377}$.

E. Lehmer showed in 1964 that there exist infinitely many Fibonacci pseudoprimes; more precisely, if p is any prime greater than
5, then U_{2p} is a Fibonacci pseudoprime.

Property (X.2) was investigated by Parberry (in 1970) and later
by Yorinaga (1976).

Among his several results, Parberry showed that if $\gcd(h, 30) = 1$
and condition (X.2) is satisfied by h, then it is also satisfied by
$k = U_h$; moreover, $\gcd(k, 30) = 1$ and, if h is composite clearly,
U_h is also composite. This shows that if there exists one composite
Fibonacci number U_n such that $U_n \equiv (5/n) \pmod{n}$, then there
exist infinitely many such numbers. As I shall say (in a short while)
there do exist such Fibonacci numbers.

Actually, this follows also from another result of Parberry:

If p is prime and $p \equiv 1$ or $4 \pmod{15}$, then $n = U_{2p}$ is odd composite and it satisfies both properties (X.1) and (X.2). In particular,
there are infinitely many Fibonacci pseudoprimes which, moreover,
satisfy (X.2). (Here I use the fact—to be indicated later in Chapter
4, Section IV—that there exist infinitely many primes p such that
$p \equiv 1 \pmod{15}$, resp. $p \equiv 4 \pmod{15}$.)

If $p \not\equiv 1$ or $4 \pmod{15}$, then (X.2) is not satisfied, as follows from
various divisibility properties and congruences indicated in Section
IV.

Yorinaga considered the primitive part of the Fibonacci number
U_n. If you remember, I have indicated in Section IV that every Fibonacci number U_n (with $n \neq 1, 2, 6, 12$) admits a primitive prime
factor p—these are the primes that divide U_n, but do not divide U_d,
for every d, $1 < d < n$, d dividing n. Thus $U_n = U_n^* \times U_n'$, where
$\gcd(U_n^*, U_n') = 1$ and p divides U_n^* if and only if p is a primitive prime
factor of U_n.

Yorinaga showed that if m divides U_n^* (with $n > 5$) then $U_m \equiv (5/m) \pmod{m}$.

According to Schinzel's result (1963), discussed in Section IV, there exist infinitely many integers n such that U_n^* is not a prime. So, Yorinaga's result implies that there exist infinitely many odd composite n such that the congruence (X.2) is satisfied.

Yorinaga published a table of all 109 composite numbers n up to 707000, such that $U_n \equiv (5/n) \pmod{n}$. Some of these numbers give also Fibonacci pseudoprimes, like $n = 4181 = 37 \times 113$, $n = 5777 = 53 \times 109$, and many more. Four of the numbers in the table give pseudoprimes in base 2:

$$
\begin{aligned}
219781 &= 271 \times 811 \\
252601 &= 41 \times 61 \times 101 \\
399001 &= 31 \times 61 \times 211 \\
512461 &= 31 \times 61 \times 271.
\end{aligned}
$$

Another result of Parberry, later generalized by Baillie & Wagstaff, is the following:

If n is an odd composite number, not a multiple of 5, if congruences (X.1) and (X.2) are satisfied, then

$$
\begin{cases}
U_{(n-(5/n))/2} \equiv 0 \pmod{n} & \text{if } n \equiv 1 \pmod{4}, \\
V_{(n-(5/n))/2} \equiv 0 \pmod{n} & \text{if } n \equiv 3 \pmod{4}.
\end{cases}
$$

In particular, since there are infinitely many composite integers n such that $n \equiv 1 \pmod{4}$, then there are infinitely many odd composite integers n satisfying the congruence $U_{(n-(5/n))/2} \equiv 0 \pmod{n}$.

The composite integers n such that $V_n \equiv 1 \pmod{n}$ [where $(V_k)_{k \geq 0}$ is the sequence of Lucas numbers] have also been studied. They have been called *Lucas pseudoprimes*, but this name is used here with a different meaning.

In 1983, Singmaster found the following 25 composite numbers $n < 10^5$ with the above property:

705, 2465, 2737, 3745, 4181, 5777, 6721,
10877, 13201, 15251, 24465, 29281, 34561,
35785, 51841, 54705, 64079, 64681, 67861,
68251, 75077, 80189, 90061, 96049, 97921.

B. LUCAS PSEUDOPRIMES ($\ell\,\mathrm{psp}(P,Q)$)

I shall now consider $\ell\mathrm{psp}(P,Q)$ associated to arbitrary pairs of parameters (P,Q). To stress the analogy with the pseudoprimes in base a, the discussion should follow the same lines. But, it will be clear that much less is known about these numbers. For example, there is no explicit mention of any algorithm to generate infinitely many $\ell\mathrm{psp}(P,Q)$, when P, Q are given—except the results mentioned for Fibonacci pseudoprimes.

However, in his thesis in 1971, Lieuwens stated that for every $k \geq 2$, there exist infinitely many Lucas pseudoprimes with given parameters (P,Q), which are the product of exactly k distinct primes.

It is quite normal for an odd integer n to be a Lucas pseudoprime with respect to many different sets of parameters. Let $D \equiv 0$ or 1 (mod 4), let $B_{\ell\,\mathrm{psp}}(n,D)$ denote the number of integers P, $1 \leq P \leq n$, such that there exists Q, with $P^2 - 4Q \equiv D \pmod{n}$ and n is a $\ell\mathrm{psp}(P,Q)$. Baillie & Wagstaff showed in 1980 that

$$B_{\ell\,\mathrm{psp}}(n,D) = \prod_{p \mid n} \left\{ \gcd\left(n - \left(\frac{D}{n}\right), p - \left(\frac{D}{p}\right)\right) - 1 \right\}.$$

In particular, if n is odd and composite, there exists D and, correspondingly, at least three pairs (P,Q), with $P^2 - 4Q = D$ and distinct values of P modulo n, such that n is a $\ell\mathrm{psp}(P,Q)$.

Another question is the following: If n is odd, for how many distinct D modulo n, do there exist (P,Q) with $P^2 - 4Q \equiv D \pmod{n}$, $P \not\equiv 0 \pmod{n}$, and n is a $\ell\mathrm{psp}(P,Q)$? Baillie & Wagstaff discussed this matter also, when $n = p_1 p_2$, where p_1, p_2 are distinct primes.

C. EULER–LUCAS PSEUDOPRIMES ($e\ell\,\mathrm{psp}(P,Q)$) AND STRONG LUCAS PSEUDOPRIMES ($s\ell\,\mathrm{psp}(P,Q)$)

Let P, Q be given, $D = P^2 - 4Q$, as before. Let n be an odd prime number. If $\gcd(n,QD) = 1$, it was seen in Section V that

$$\begin{cases} U_{(n-(D/n))/2} \equiv 0 \pmod{n} & \text{when } (Q/n) = 1, \\ V_{(n-(D/n))/2} \equiv D \pmod{n} & \text{when } (Q/n) = -1. \end{cases}$$

This leads to the following definition. An odd composite integer n, such that $\gcd(n,QD) = 1$, satisfying the above condition is called a *Euler–Lucas pseudoprime with parameters* (P,Q), abbreviated $e\ell\mathrm{psp}(P,Q)$.

Let n be an odd composite integer, with $\gcd(n, D) = 1$, let $n - (D/n) = 2^s d$, with d odd, $s \geq 0$. If

$$\begin{cases} U_d \equiv 0 \pmod{n}, & \text{or} \\ V_{2^r d} \equiv 0 \pmod{n}, & \text{for some } r, 0 \leq r < s, \end{cases}$$

then n is called a *strong Lucas pseudoprime with parameters* (P, Q), abbreviated sℓpsp(P, Q). In this case, necessarily, $\gcd(n, Q) = 1$.

If n is an odd prime, and $\gcd(n, QD) = 1$, then n is an eℓpsp(P, Q) and also a sℓpsp(P, Q). It is also clear that if n is an eℓpsp(P, Q) and $\gcd(n, Q) = 1$, then n is a ℓpsp(P, Q).

What are the relations between eℓpsp(P, Q) and sℓpsp(P, Q)? Just as in the case of Euler and strong pseudoprimes in base a, Baillie & Wagstaff showed that if n is a sℓpsp(P, Q), then n is an eℓpsp(P, Q)—this is the analogue of Selfridge's result.

Conversely, if n is an eℓpsp(P, Q) and either $(Q/n) = -1$ or $n - (D/n) \equiv 2 \pmod{4}$, then n is a sℓpsp(P, Q)—this is the analogue of Malm's result.

If $\gcd(n, Q) = 1$, n is a ℓpsp(P, Q), $U_n \equiv (D/n) \pmod{n}$ and if, moreover, n is an epsp(Q), then n is also a sℓpsp(P, Q).

The special case for Fibonacci numbers was proved by Parberry, as already indicated.

Previously, I mentioned the result of Lehmer, saying that no odd composite number can be an epsp(a), for all possible bases.

Here is the analogous result of Williams (1977):

Given $D \equiv 0$ or $1 \pmod{4}$, if n is an odd composite integer, and $\gcd(n, D) = 1$, there exist P, Q, nonzero integers, with $P^2 - 4Q = D$, $\gcd(P, Q) = 1$, $\gcd(n, Q) = 1$, and such that n is not an eℓpsp(P, Q).

With the present terminology, I have mentioned already that Parberry had shown, for the Fibonacci sequence, that there exist infinitely many eℓpsp$(1, -1)$.

This has been improved by Kiss, Phong & Lieuwens (1986): Given (P, Q) such that the sequence $(U_n)_{n \geq 0}$ is nondegenerate (that is, $U_n \neq 0$ for every $n \neq 0$), given $k \geq 2$, there exist infinitely many eℓpsp(P, Q), each being the product of k distinct primes. Moreover, given also $d \geq 2$, if $D = P^2 - 4Q > 0$, then the prime factors may all be chosen to be of the form $dm + 1 (m \geq 1)$.

As for Fibonacci numbers, now I consider the congruences (X.2) and also (X.3), (X.4).

It may be shown that if $\gcd(n, 2PQD) = 1$ and if n satisfies any two of the congruences (X.1) to (X.4), then it satisfies the other two.

In 1986, Kiss, Phong & Lieuwens extended a result of Rotkiewicz (1973) and proved:

Given, P, $Q = \pm 1$ [but $(P, Q) \neq (1, 1)$], given $k \geq 2$, $d \geq 2$, there exist infinitely many integers n, which are Euler pseudoprimes in base 2, and which satisfy the congruences (X.1)–(X.4); moreover, each such number n is the product of exactly k distinct primes, all of the form $dm + 1$ (with $m \geq 1$).

D. CARMICHAEL–LUCAS NUMBERS

Following the same line of thought that led from pseudoprimes to Carmichael numbers, it is natural to consider the following numbers.

Given $D \equiv 0$ or $1 \pmod 4$, the integer n is called a *Carmichael–Lucas number* (*associated to* D), if $\gcd(n, D) = 1$ and for all nonzero relatively prime integers P, Q with $P^2 - 4Q = D$ and $\gcd(n, Q) = 1$, the number is an $\ell\mathrm{psp}(P, Q)$.

Do such numbers exist? A priori, this is not clear. Of course, if n is a Carmichael–Lucas number associated to $D = 1$, then n is a Carmichael number.

Williams, who began the consideration of Carmichael–Lucas numbers, showed in (1977):

If n is a Carmichael–Lucas number associated to D, then n is the product of $k \geq 2$ distinct primes p_i such that $p_i - (D/p_i)$ divides $n - (D/n)$.

Note that $323 = 17 \times 19$ is a Carmichael—Lucas number (with $D = 5$); but it cannot be a Carmichael number, because it is the product of only two distinct primes.

Adapting the method of Chernick, it is possible to generate many Carmichael–Lucas numbers. Thus, for example, $1649339 = 67 \times 103 \times 239$ is such a number (with $D = 8$).

XI. Last Section on Primality Testing and Factorization!

Only one section—and a small one—to treat a burning topic, full of tantalizing ideas and the object of intense research, in view of immediate direct applications.

Immediate direct applications of number theory! Who would dream of it, even some 30 years ago? Von Neumann yes, not me, not many people. Poor number theory, the Queen relegated (or raised?) to be the object of a courtship inspired by necessity not by awe.

In recent years, progress on the problems of primality testing and factorization has been swift. More and more deep results of number theory have been invoked. Brilliant brains devised clever procedures, not less brilliant technicians invented tricks, shortcuts to implement the methods in a reasonable time—and thus, a whole new branch of number theory is evolving.

I shall present here an overview of certain questions related to primality and factorization. The latest developments in this area require deep number-theoretic results which cannot be described in this book. Happily enough, there are good expository articles, which I will recommend at the right moment.

First, money: how much it costs to perform the tests. Then, I shall discuss more amply primality tests, indicate some noteworthy recent factorizations, to conclude with a quick description of applications to public key cryptography.

A. MONEY

The cost of performing an algorithm depends on the time required to arrive at the output. The speed of the computer is a factor, which is taken care by an appropriate constant. The time is then proportional to the number of operations, or more precisely, to the number of operations on the digits of the given number N; thus the input is taken to be $\log N$, which is proportional to the number of digits of N.

The algorithm runs in *polynomial time* if there exists a polynomial $f(x)$ such that, for every N, the time required to perform the algorithm on the number N is bounded by $f(\log N)$. An algorithm, not of polynomial time, whose running time is bounded by $f(N)$ (for every N) where $f(x)$ is a polynomial, is said to have an *exponential* running time, since $N = e^{\log N}$. An algorithm can only be economically justified if it runs in polynomial time.

The theory of complexity of algorithms deals specifically with the determination of bounds for the running time. It is a very elaborate sort of bookkeeping, which requires a careful analysis of the methods involved. Through the discovery of clever tricks, algorithms may sometimes be simplified into others requiring only a polynomial running time.

It may be said that the main problem faced in respect to primality testing (and many other problems) is the following:

Does there exist an algorithm to perform the test, which runs in polynomial time?

For primality testing, this is an open problem. On the one hand, it has not been shown that such a polynomial time algorithm cannot exist. On the other hand, nobody has found one!

All this should not be confused with the following.

If a number N is known to be composite, this fact may be proved with only one operation. Indeed, it is enough to produce two numbers a, b, such that $N = ab$, so the number of bit operations required is at most $(\log N)^2$. Paraphrasing Lenstra, it is irrelevant whether a, b were found after consulting a clairvoyant, or after three years of Sundays, like Cole's factorization of the Mersenne number M_{67}:

$$2^{67} - 1 = 193707721 \times 761838257287.$$

If N is known to be a prime, what is the number of bit operations required to prove it? This is not so easy to answer. In 1975, Pratt showed that it suffices a $C(\log N)^4$ bit operations (where C is a positive constant).

B. More Primality Tests

I return once more to primality testing. There are many kinds, and according to the point of view, these tests may be classified as follows:

$$\begin{cases} \text{Tests for numbers of special forms} \\ \text{Tests for generic numbers} \end{cases}$$

or

$$\begin{cases} \text{Tests with full justification} \\ \text{Tests with justification based on conjectures} \end{cases}$$

or

$$\begin{cases} \text{Deterministic tests} \\ \text{Probabilistic or Monte Carlo tests.} \end{cases}$$

In the sequel, I shall encounter tests of each of the above kinds.

If sufficiently many prime factors of $N - 1$ or $N + 1$ are known, the tests indicated in Sections III and V are applicable. It is easy to show that the running time of these tests is at most $C(\log N)^3$. Thus, these tests are of polynomial running time.

For numbers that are not of a special form, the very naive primality test is by trial division of N by all primes $p < \sqrt{N}$. It will be seen in Chapter 4 that, for any large integer N, the number of primes less than N is about $N/(\log N)$ (this statement will be made much more precise later on); thus there will be at most $C[\sqrt{N}/(\log N)]$ operations, which tells that the running time is $C\sqrt{N} \log N$. So this procedure does not run in polynomial time on the input.

MILLER'S TEST

In 1976, Miller proposed a primality test, which was justified using a generalized form of Riemann's hypothesis. I will not explain the exact meaning of this hypothesis or conjecture, but in Chapter 4, I shall discuss the classical Riemann's hypothesis.

To formulate Miller's test, which involves the congruences used in the definition of strong pseudoprimes, it is convenient to use the terminology introduced by Rabin.

If N is an integer, if $1 < a < N$, if $N - 1 = 2^s d$, with $s \geq 0$, d odd, a is said to be a *witness* for N when $a^{N-1} \not\equiv 1 \pmod{N}$ or there exists r, $0 \leq r < s$ such that $1 < \gcd(a^{2^r d} - 1, N) < N$.

If N has a witness, it is composite.

If N is composite, if $1 < a < N$, and a is not a witness, then $\gcd(a, N) = 1$ and N is a spsp(a). Conversely, if N is odd and N is a spsp(a) then a is not a witness for N.

In the new terminology, in order to certify that N is prime, it suffices to show that no integer a, $1 < a < N$, $\gcd(a, N) = 1$, is a witness. Since N is assumed to be very large, this task is overwhelming! It would be wonderful just to settle the matter by considering small integers a, and checking whether any one is a witness for N. Here is where the generalized Riemann's hypothesis is needed: to show that, it suffices to consider much smaller bases a:

Miller's Test. Let N be an odd integer. If there exists a, such that $\gcd(a, N) = 1$, $1 < a < 2(\log N)^2$, which is a witness for N, then N is "guilty," that is, composite. Otherwise, N is "innocent," that is, a prime.

I should add here that for numbers up to 25×10^9, because of the calculations reported in Section VIII, the only composite integers $N < 3 \times 10^9$ that are strong pseudoprimes simultaneously to the bases 2, 3, 5 are the four numbers 25,326,001; 161,304,001; 960,946,321; 1,157,839,381. So, if N is not one of these numbers and 2, 3, 5 are not witnesses for N, then N is a prime. Similarly, 3,215,031,751 is the only number $N < 25 \times 10^9$ which is a strong pseudoprime for the bases 2, 3, 5, 7. So if $N < 25 \times 10^9$, N not the above number, and 2, 3, 5, 7 are not witnesses, then N is a prime.

This test may be easily implemented on a pocket calculator.

The number of bit operations for testing whether a number is a witness for N is $C(\log N)^5$. So, this test runs in polynomial time on the input (provided the generalized Riemann's hypothesis is assumed true).

Miller's test is discussed in Lenstra's paper (1982) as well as in the nice recent expository paper of Wagon (1986).

The APR Test

The primality test devised by Adleman, Pomerance & Rumely (1983), usually called the APR test, represents a breakthrough. To wit:

(i) It is applicable to arbitrary natural numbers N, without requiring the knowledge of factors of $N - 1$ or $N + 1$.

(ii) The running time $t(N)$ is almost polynomial: more precisely, there exist effectively computable constants $0 < C' < C$, such that

$$(\log N)^{C' \log \log \log N} \le t(N) \le (\log N)^{C \log \log \log N}.$$

(iii) The test is justified rigorously, and for the first time ever in this domain, it is necessary to appeal to deep results in the theory of algebraic numbers; it involves calculations with roots of unity and the general reciprocity law for the power residue symbol. (Did you notice that I have not explained these concepts? It is far beyond what I plan to treat.)

The running time of the APR is at the present the world record for a deterministic primality test.

Soon afterwards, Cohen & Lenstra (1984) modified the APR test, making it more flexible, using Gauss sums in the proof (instead of the reciprocity law), and having the new test programmed for practical applications. It was the first primality test in existence that can routinely handle numbers of up to 100 decimal digits, and it does so in approximately 45 seconds.

A presentation of the APR test was made by Lenstra in the Séminaire Bourbaki, exposé 576 (1981). It was also discussed in papers of Lenstra (1982) and Nicolas (1984).

This performance has been surpassed. Cohen & Lenstra, Br. (Brother, not Junior), tested a number of 213 digits for primality, in about ten seconds (1987). And by the time you will be reading this book, this information will be outdated.

MONTE CARLO METHODS

Early in this century, the casino in Monte Carlo attracted aristocracy and adventurers, who were addicted to gambling. Tragedy and fortune were determined by the spinning wheel.

I read with particular pleasure the novel by Luigi Pirandello, telling how the life of Mattia Pascal was changed when luck favored him, both at Monte Carlo and in his own Sicilian village. When a corpse was identified as being his, he could escape the inferno of his married life, and became RICH and FREE! But Monte Carlo is not always so good. More often, total ruin, followed by suicide, is the price paid.

As you will enter into the Monte Carlo primality game, if your Monte Carlo testing will be unsuccessful, I sincerely hope that you will not be driven to suicide.

I wish to mention three Monte Carlo tests, due to Baillie & Wagstaff (1980), Solovay & Strassen (1977) and Rabin (1976, 1980). In each of these tests a number of witnesses a are used, in connection with congruences like those satisfied by psp (a), epsp (a), spsp(a).

I describe briefly Rabin's test, which is very similar to Miller's.

Based on the same idea of Solovay & Strassen, Rabin proposed the following test:

Step 1. Choose, at random, $k > 1$ small numbers a, such that $1 < a < N$, $\gcd(a, N) = 1$.

Step 2. Test, in succession, for each chosen basis a, whether N satisfies the condition in the definition of a strong pseudoprime in base a; writing $N - 1 = 2^s d$, with d odd, $s \geq 0$, either $a^d \equiv 1 \pmod{N}$ or $a^{2^r d} \equiv -1 \pmod{N}$ for some r, $0 \leq r < s$.

If an a is found for which the above condition does not hold, then declare N to be composite. Otherwise, the probability that N is a prime, when certified prime, is at least $1 - 1/4^k$. So, for $k = 30$, the likely error is at most one in 10^{18} tests.

You may wish to sell prime numbers—yes, I say sell—to be used in public key cryptography (be patient, I will soon come to this application of primality and factorization). And you wish to be sure, or sure with only a negligible margin of error, that you are really selling a prime number, so that you may advertise: "Satisfaction guaranteed or money back."

On the basis of Rabin's test, you can safely develop a business and honestly back the product sold.

C. TITANIC PRIMES

In an article of 1983/4, Yates coined the expression "titanic prime" to name any prime with at least 1000 digits. In the paper, with the suggestive title *"Sinkers of the Titanics"* (1984/5), Yates compiled a list of the 110 largest known titanic primes. By January 1, 1985, he knew 581 titanic primes, of which 170 have more than 2000 digits.

In September 1988, Yates' list comprised already 876 titanic primes; the "Six of Amdahl" announced, early 1990, 550 new titanic primes. . . .

With the extraordinary energy of the sinkers, it will not be long before the list of titanic primes becomes larger than the total number of lines in this book. No more bad conscience for not reporting these efforts. . . . However, I would not be forgiven for ignoring the following curiosities.

The largest prime all of whose digits are prime numbers is

$$7532 \times \frac{10^{1104} - 1}{10^4 - 1} + 1 :$$

it has 1104 digits and was discovered by Dubner, in 1988.

The largest prime with one digit 1 and all other digits equal to 9 is $2 \times 10^{3020} - 1$; it has 3021 digits and was discovered by Williams in 1985.

The largest prime with one digit 5 and all other digits equal to 9, is $6 \times 10^{4332} - 1$; it has 4333 digits, and was discovered by Dubner in 1988.

The largest prime all of whose digits are odd is $1358 \times 10^{3821} - 1$; it has 3825 digits and was discovered by Dubner in 1988.

The largest prime with all digits equal to 0 or to 1, is

$$\frac{10^{640} - 1}{9} \times 10^{641} + 1;$$

it has 1281 digits and was discovered by Dubner in 1984.

And last (but surely least!), the smallest known titanic prime is $25 \times 2^{3314} - 1$, with exactly 1000 digits, discovered by Cormack & Williams in 1979.

It is not surprising that these primes have special forms, many being Mersenne primes, others being of the form $k \times 2^n \pm 1$, $k^2 \times 2^n + 1$, $k^4 \times 2^n + 1$, $k \times 10^n + 1$, $(10^n - 1)/9$.

Of course, it is for numbers of these forms that the more efficient primality testing algorithms are available.

Lenstra published in 1987 a paper about factorization using elliptic curves; the paper of the brothers Lenstra of 1989 is also of fundamental importance. More recently, there is the preprint on the number field sieve, by the brothers Lenstra, Manasse & Pollard (1989).

The recent volume by Brillhart, Lehmer, Selfridge, Tuckerman & Wagstaff (1983; second edition, 1988) contains tables of known factors of $b^n \pm 1$ ($b = 2$ 3, 5, 6, 7, 10, 11, 12) for various ranges of n. For example, the table of factors of $2^n - 1$ extends for $n < 1200$; for larger bases b, the range is smaller. The second edition of the book, contains 2045 new factorizations, reflecting the important progress accomplished in the last few years, both in the methods and in the technology. All ten "most wanted" factorizations of the first edition have now been performed. To wit, it was known that

$$2^{211} - 1 = 15193 \times C\ 60$$

where $C\ 60$ indicates a number with 60 digits—the first "most wanted" number to be factored. This has now been done.

At the origin, this collective work, also dubbed "the Cunningham project", was undertaken to extend the tables published by Cunningham & Woodall in 1925.

D. FACTORIZATION

The factorization of large integers is a hard problem: there is no known algorithm that runs in polynomial time. It is also an important problem, because it has found a notorious application to public key cryptography.

Nevertheless, I shall not discuss here the methods of factorization— this would once again lead me too far from the subject of records on prime numbers.

Here are a few indications in the literature.

The expository paper of Guy (1975) discusses the methods, now considered classical; Williams (1984) covers about the same ground, being naturally more up to date. The lecture notes of a short course by Pomerance (1984) contains an annotated bibliography. Dixon wrote well in 1984 about factorization and primality.

Riesel published in 1985 a book devoted to factorization, which discusses in a lucid way the methods available at the time. It is a good place to study techniques of factorization which are exposed in a coherent and unified way.

It also contains tables of factors of Fermat numbers, of Mersenne numbers, of numbers of the forms $2^n + 1$, $10^n + 1$, or repunits $(10^n - 1)/9$, and many more.

Just as an illustration, and for the delight of lovers of large numbers, I will give now explicit factorizations of some Mersenne, Fermat, and other numbers [for the older references, see Dickson's *History of the Theory of Numbers*, vol. I, pages 22, 29, 377, and Archibald (1935)].

$M_{59} = 2^{59} - 1 = (179951) \times (3203431780337)$, by Landry in 1869;

$M_{67} = 2^{67} - 1 = (193707721) \times (761838257287)$, by Cole in 1903, already indicated;

$M_{73} = 2^{73} - 1 = (439) \times (2298041) \times (9361973132609)$, the factor 439 by Euler, the other factors by Poulet in 1923;

$F_6 = 2^{2^6} + 1 = (1071 \times 2^8 + 1) \times (262814145745 \times 2^8 + 1) = (274177) \times (67280421310721)$, by Landry in 1880.

The above factorizations were obtained before the advent of computers!

More recently, the following factorizations were obtained:

$M_{113} = 2^{113} - 1 = (3391) \times (23279) \times (65993) \times (1868569) \times (1066818132868207)$, the smallest factor by Reuschle (1856) and the remaining factors by Lehmer in 1947;

$M_{193} = 2^{193} - 1 = (13821503) \times (61654440233248340616559) \times (14732265321145317331353282383)$, by Pomerance & Wagstaff in 1983;

$M_{257} = 2^{257} - 1 = (535006138814359) \times (1155685395246619182673033) \times (374550598501810936581776630096313181393)$, by Penk & Baillie (1979, 1980); this number had been certified to be composite by Lehmer, already in 1927, but no factor had been indicated.

$F_7 = 2^{2^7} + 1 = (59649589127497217) \times$
(5704689200685129054721), by Morrison & Brillhart (in 1970, published in 1971);

$F_8 = 2^{2^8} + 1 = (1238926361552897) \times$
(93461639715357977769163558199606896584051237541638188580280321),

by Brent & Pollard in 1980;

The Fermat number F_{11} has been completely factored in 1989. Two small prime factors were long well-known; two more prime factors were found by Brent (with the elliptic curve method), who indicated that the 564-digit cofactor was probably a prime; this was shown to be the case by F. Morain.

The most recently completely factored Fermat number is F_9. It could not resist the number field sieve method (1990, by A. K. Lenstra & M. S. Manasse).

In a paper of 1988, dedicated to Dov Jarden, Brillhart, Montgomery & Silverman, give the known factors of Fibonacci numbers U_n (for $n \leq 999$) and of Lucas numbers V_n(for $n \leq 500$). This pushes much further the work which had been done by many other numerologists, among whom Jarden (see the 3rd edition of his book, 1958).

Here are some more noteworthy factorizations, which at their time represented a step forward:

$(10^{103}+1)/11 = (1237) \times (44092859) \times (102860539) \times (984385009) \times$
$(612053256358933) \times (18272511486652115647161) \times$
(1471865453993855302660887614137521979),

factorization completed by Atkin & Rickert, in 1984.

A. K. Lenstra and M. S. Manasse were "pleased to announce the first factorization of a 100-digit number by a general purpose factorization algorithm" (October 12, 1988); such an algorithm factors a number N in a deterministic way, based solely on the size of N, and not on any particular property of its factors; in its worst case, the running time for factorization is the same as the average running time.

The happy number was

$$\frac{11^{104} + 1}{11^8 + 1} =$$

86759222313428390812218077098507080489 77 ×
10848810485363747061296139984297294840983461152579057 72116753.

The number field sieve method was used to completely factor the 138-digits number $2^{457} + 1$, which is equal to $3 \times (P49) \times (P89)$. This was accomplished by A. K. Lenstra & M. S. Manasse in November 1989; newspapers reported this feat, sometimes at front page!

The second edition of the book by Brillhart et al. on the Cunningham project contains also many, many new factorizations of large numbers.

It is likely that this activity will go on unabated. Heaven is the limit!...

Anyone interested in primality testing, factorization, or similar calculations with very large numbers needs, of course, access to high-speed sophisticated computers of the latest generation. There is still pioneering work to be done in the development of gadgets adaptable to personal computers. These will allow us to reach substantial results in the comfort of home. If it is snowing outside—as is often the case in Canada—you may test your prime, keeping warm feet.

E. Public Key Cryptography

Owing to the proliferation of means of communication and the need of sending messages—like bank transfers, love letters, instructions for buying stocks, secret diplomatic informations, as, for example, reports of spying activities—it has become very desirable to develop a safe method of coding messages. In the past, codes have been kept secret, known only to the parties sending and receiving the messages. But, it has been often possible to study the intercepted messages and crack the code. In war situations, this had disastrous consequences.

Great progress in cryptography came with the advent of public key crypto-systems.

The main characteristics of the system are its simplicity, the public key, and the extreme difficulty to crack it. The idea was proposed in 1976 by Diffie & Hellman, and the effective implementation was achieved in 1978 by Rivest, Shamir, & Adleman. This crypto-system is therefore called the RSA-system. I shall describe it now.

Each letter or sign, including blank space, corresponds to a 3-digit number. In the American Standard Code for Information Interchange, this correspondence is the following:

—	A	B	C	D	E	F	G	H
032	065	066	067	068	069	070	071	072

I	J	K	L	M	N	O	P	Q
073	074	075	076	077	078	079	080	081

R	S	T	U	V	W	X	Y	Z
082	083	084	085	086	087	088	089	090

Each letter or sign of a message is replaced by the corresponding 3-digit number, giving rise to a number, M which represents the actual message.

Thus a message M may be viewed as a positive integer, namely, writing the string of triple digit numbers corresponding to the letters and signs of the message.

Each user A of the system lists in a public directory his key, which is a pair of positive integers: (n_A, s_A). The first integer n_A is a product of two primes p_A, q_A, $(n_A = p_A q_A)$, which are chosen to be large and are kept secret. Moreover, s_A is chosen to be relatively prime with both $p_A - 1$, $q_A - 1$.

To send a message M to another user B, A encrypts M—the way to encode M depends on who will receive it. Upon receiving the encoded message from A, the user B decodes it using his own secret decoding method.

In detail, the process goes as follows. If the message $M \geq n_B$, it suffices to break M into smaller blocks; so it may be assumed that $M < n_B$. If $\gcd(M, n_B) \neq 1$, a dummy letter is added to the end of M, so that for the new message, $\gcd(M, n_B) = 1$.

A sends to B the encoded message $E_B(M) = M'$, $1 \leq M' < n_B$, where M' is the residue of M^{s_B} modulo n_B: $M' \equiv M^{s_B} \pmod{n_B}$.

In order to decode M', the user B calculates t_B, $1 \leq t_B < (p_B - 1)(q_B - 1) = \phi(n_B)$, such that $t_B s_B \equiv 1 \pmod{\phi(n_B)}$; this is done once and for all. Then

$$D_B(M') = M'^{t_B} \equiv M^{s_B t_B} \equiv M \pmod{n_B},$$

so B may read the message M.

How simple!

Put your hand in your pocket and pick your little calculator. We shall play a game, which for many of you will also require a dictionary.

Below is a message which a certain person is sending to an individual whose public key is (m, s), where $m = 156287$, $s = 181$:

1522140146850141731113220073001378790850700146850 10359

1079071542451005410383140508830898930757060619490 18136

1079070365821391480046060530691214640634920054910 19712

1031031150790543440469951276930103590827750695920 23529

0383140508830852020371280507060854730952750415210 01124

1031030136590727530029521145860 72755

You don't know the secret prime factors of m. Can you decode the message?

Do you understand what is says? Who sent the message?

Since I am very helpful, I shall now say a little bit on how to crack a crypto-system. It is necessary to discover $\phi(n_A)$ for each user A.

This is equivalent to the factorization of n_A. Indeed, if p_A, q_A are known, then $\phi(n_A) = (p_A - 1)(q_A - 1)$. Conversely, putting $p = p_A$, $q = q_A$, $n = n_A$, from $\phi(n) = (p - 1)(q - 1) = n + 1 - (p + q)$, $(p + q)^2 - 4n = (p - q)^2$ (if $p > q$), then

$$
\begin{aligned}
p + q &= n + 1 - \phi(n), \\
p - q &= \sqrt{[n + 1 - \phi(n)]^2 - 4n},
\end{aligned}
$$

and from this, p, q are expressed in terms of n, $\phi(n)$.

If p, q are large primes, say with 100 or more digits, and chosen at random, the factorization of n, with 200 or more digits, is, in general, unfeasible with the methods known today.

There is much more to be said about the RSA crypto-system:

a) how to send "signed" messages, so that the receiver can unmistakably identify the sender.

b) how to choose well the prime factors of the numbers n_A of the keys, so that the cracking of the system is less probable.

The reader may consult the original papers of Rivest, Shamir & Adleman (1978), of Rivest (1978), as well as the books of Riesel (1985), Koblitz (1987). And for example the lecture notes of Lemos (1989), which are written in Portuguese—it is like studying cryptography in an encrypted language. Perhaps all this in Copacabana Beach.

3

Are There Functions Defining Prime Numbers?

To determine prime numbers, it is natural to ask for functions $f(n)$ defined for all natural numbers $n \geq 1$, which are computable in practice and produce some or all prime numbers.

For example, one of the following conditions should be satisfied:

(a) $f(n) = p_n$ for every $n \geq 1$;

(b) $f(n)$ is always a prime number, and if $n \neq m$, then $f(n) \neq f(m)$;

(c) the set of prime numbers is equal to the set of positive values assumed by the function.

Clearly, condition (a) is more demanding than (b) and than (c).

The results in this direction have been rather disappointing, except for some of theoretical importance, related to condition (c).

I. Functions Satisfying Condition (a)

The result of Willans (1964) is somewhat interesting, even though of no practical use. It provides definite answers to Hardy & Wright's questions in their famous book:

(1) Is there a formula for the nth prime number?

(2) Is there a formula for a prime, in terms of the preceding primes?

What is intended here is to find a closed expression for the nth prime p_n, in terms of n, by means of functions that are computable and, if possible, classical. Intimately related with this problem is to find reasonable expressions for the function counting primes.

For every real number $x > 0$ let $\pi(x)$ denote the number of primes p such that $p \leq x$.

This is a traditional notation for one of the most important functions in the theory of prime numbers. I shall return to it in Chapter 4. Even though the number $\pi = 3.14\ldots$ and the function $\pi(x)$ do occur below in the same formula, this does not lead to any ambiguity.

First I indicate a formula for $\pi(m)$, given by Willans. It is based on the classical Wilson's theorem which I proved in Chapter 2. We also indicate by $[x]$ the only integer n such that the real number x verifies $n \leq x < n + 1$.

For every integer $j \geq 1$ let

$$F(j) = \left[\cos^2 \pi \frac{(j-1)! + 1}{j}\right].$$

So for any integer $j > 1$, $F(j) = 1$ when j is a prime, while $F(j) = 0$ otherwise. Also $F(1) = 1$.

Thus

$$\pi(m) = -1 + \sum_{j=1}^{m} F(j).$$

Willans also expressed:

$$\pi(m) = \sum_{j=2}^{m} H(j) \qquad \text{for } m = 2, 3, \ldots$$

where

$$H(j) = \frac{\sin^2 \pi \frac{\{(j-1)!\}^2}{j}}{\sin^2\left(\frac{\pi}{j}\right)}$$

Mináč gave an alternate (unpublished) expression, which involves neither the cosine nor the sine:

$$\pi(m) = \sum_{j=2}^{m} \left[\frac{(j-1)! + 1}{j} - \left[\frac{(j-1)!}{j}\right]\right].$$

Proof. The proof of Mináč's formula is quite simple, and since it is not published anywhere, I'll give it here.

First a remark: if $n \neq 4$ is not a prime, then n divides $(n-1)!$. Indeed, either n is equal to a product $n = ab$, with $2 \leq a$, $b \leq n-1$ and $a \neq b$, or $n = p^2 \neq 4$. In the first alternative, n divides $(n-1)!$; in the second case, $2 < p \leq n - 1 = p^2 - 1$, so $2p \leq p^2 - 1$ and n divides $2p^2 = p \times 2p$, which divides $(n-1)!$.

For any prime j, by Wilson's theorem, $(j-1)! + 1 = kj$ (where k is an integer), so

$$\left[\frac{(j-1)!+1}{j} - \left[\frac{(j-1)!}{j}\right]\right] = \left[k - \left[k - \frac{1}{j}\right]\right] = 1.$$

If j is not a prime and $j \geq 6$, then $(j-1)! = kj$ (where k is an integer), by the above remark. Hence

$$\left[\frac{(j-1)!+1}{j} - \left[\frac{(j-1)!}{j}\right]\right] = \left[k + \frac{1}{j} - k\right] = 0.$$

Finally, if $j = 4$, then

$$\left[\frac{3!+1}{4} - \left[\frac{3!}{4}\right]\right] = 0.$$

This is enough to prove the formula indicated for $\pi(m)$.   □

With the above notations, Willans gave the following formula for the nth prime:

$$p_n = 1 + \sum_{m=1}^{2^n} \left[\left(\frac{n}{\sum_{j=1}^{m} F(j)}\right)^{1/n}\right]$$

or, in terms of the prime counting function:

$$p_n = 1 + \sum_{m=1}^{2^n} \left[\left(\frac{n}{1 + \pi(m)}\right)^{1/n}\right].$$

For the related problem of expressing a prime q in terms of the prime p immediately preceding it, Willans gave the formula:

$$q = 1 + p + F(p+1) + F(p+1)F(p+2) + \cdots + \prod_{j=1}^{p} F(p+j),$$

where $F(j)$ was defined above.

Another formula for the smallest prime greater than $m \geq 2$ was given by Ernvall, while still a student, and published in 1975: Let

$$d = \gcd((m!)^{m!} - 1, (2m)!),$$

let

$$t = \frac{d^d}{\gcd(d^d, d!)}$$

and let a be the unique integer such that d^a divides t, but d^{a+1} does not divide t. Then the smallest prime larger than m is

$$p = \frac{d}{\gcd(t/d^a, d)}.$$

Taking $m = p_{n-1}$, this yields a formula for p_n.

In 1971, Gandhi gave a formula for the nth prime p_n. To explain it, I require the Möbius function, which is one of the most important arithmetic functions. The Möbius function is defined as follows:

$$\begin{cases} \mu(1) = 1, \\ \text{if } n \text{ is the product of r distinct primes, then } \mu(n) = (-1)^r, \\ \text{if the square of a prime divides } n, \text{ then } \mu(n) = 0. \end{cases}$$

Let $P_{n-1} = p_1 p_2 \cdots p_{n-1}$. Then Gandhi showed:

$$p_n = \left[1 - \frac{1}{\log 2} \log \left(-\frac{1}{2} + \sum_{d \mid P_{n-1}} \frac{\mu(d)}{2^d - 1} \right) \right]$$

or equivalently, p_n is the only integer such that

$$1 < 2^{p_n} \left(-\frac{1}{2} + \sum_{d \mid P_{n-1}} \frac{\mu(d)}{2^d - 1} \right) < 2.$$

The following simple proof was given by Vanden Eynden in 1972.

Proof. For simplicity of notation, let $Q = P_{n-1}$, $p_n = p$, and

$$S = \sum_{d \mid Q} \frac{\mu(d)}{2^d - 1}.$$

So

$$(2^Q - 1)S = \sum_{d \mid Q} \mu(d) \frac{2^Q - 1}{2^d - 1} = \sum_{d \mid Q} \mu(d)(1 + 2^d + 2^{2d} + \cdots + 2^{Q-d}).$$

If $0 \leq t < Q$, the term $\mu(d)2^t$ occurs exactly when d divides $\gcd(t, Q)$. So the coefficient of 2^t in the last sum is $\sum_{d \mid \gcd(t,Q)} \mu(d)$; in particular, for $t = 0$ it is equal to $\sum_{d \mid Q} \mu(d)$.

But, for any integer $m \geq 1$ it is well known and easy to show, that

$$\sum_{d|m} \mu(d) = \begin{cases} 1 & \text{if } m = 1, \\ 0 & \text{if } m > 1. \end{cases}$$

Writing $\sum'_{0<t<Q}$ for the sum extended over all t, such that $0 < t < Q$ and $\gcd(t, Q) = 1$, then $(2^Q - 1)S = \sum'_{0<t<Q} 2^t$; the largest index t in this summation is $t = Q - 1$. It follows that

$$2(2^Q - 1)\left(-\frac{1}{2} + S\right) = -(2^Q - 1) + \sum_{0<t<Q}' 2^{t+1} = 1 + \sum_{0<t<Q-1}' 2^{t+1}$$

If $2 \leq j < p_n = p$, there exist some prime q such that $q < p_n = p$ (so $q|Q$) and $q|Q - j$. Hence every index t in the above sum satisfies $0 < t \leq Q - p$. Thus

$$\frac{2^{Q-p+1}}{2 \times 2^Q} < -\frac{1}{2} + S = \frac{1 + \sum'_{0<t\leq Q-p} 2^{t+1}}{2(2^Q - 1)} < \frac{2^{Q-p+2}}{2 \times 2^Q},$$

where the inequalities are easy to establish.

Hence multiplying with 2^p, it follows that

$$1 < 2^p \left(-\frac{1}{2} + S\right) < 2.$$

$\square$

In 1974, Golomb gave another proof, based on a totally different idea, involving probability and binary expansions.

II. Functions Satisfying Condition (b)

The number $f(n) = [\theta^{3^n}]$ is a prime for every $n \geq 1$; here θ is a number which is roughly equal to $1.3064\ldots$ (see Mills, 1947). Similarly,

$$g(n) = \left[2^{2^{2^{\cdot^{\cdot^{\cdot^{2^\omega}}}}}}\right]$$

(a string of n exponents) is a prime for every $n \geq 1$; here ω is a number which is roughly equal to $1.9287800\ldots$ (see Wright, 1951).

The fact that θ, ω are known only approximately and the numbers grow very fast, make these formulas no more than curiosities. For example, $g(1) = 3$, $g(2) = 13$, $g(3) = 16381$, $g(4)$ has more than 5000 digits.

There are many other formulas of a similar kind in the literature, but they are just as useless; see Dudley (1969).

At this point one might wonder: Why not try some polynomial with integral coefficients instead of these awkward functions involving exponentials and the largest integer function?

The reason is simply given by the following negative result:

If f is a polynomial with integral coefficients, in one indeterminate, which is not constant, then there exist infinitely many integers n such that $|f(n)|$ is not a prime number.

Proof. I may assume that there exists some integer $n_0 \geq 0$ such that $|f(n_0)| = p$ is a prime number. Since the polynomial is not constant, $\lim_{x \to \infty} |f(x)| = \infty$, so there exists $n_1 > n_0$ such that if $n \geq n_1$, then $|f(n)| > p$. For any h such that $n_0 + ph \geq n_1$, $f(n_0 + ph) = f(n_0) +$ multiple of $p =$ multiple of p. Since $|f(n_0 + ph)| > p$, then $|f(n_0 + ph)|$ is a composite integer.   □

Now that no polynomial with integral coefficients in one indeterminate is fit for the purpose, could a polynomial with several indeterminates be suitable?

Once more, this is excluded by the following strong negative result:

If f is a polynomial with complex coefficients in m indeterminates such that its values at natural numbers are, in absolute value, prime numbers, then f must be a constant.

Even though nonconstant polynomials $f(X)$ with integral coefficients have composite values (in absolute value) at infinitely many natural numbers, Euler discovered in 1772 one such polynomial $f(X)$ with a "long string" of prime values. For these polynomials, there exist natural numbers m, n with $0 \leq m < n$ (and $n - m$ not too small) such that $f(k)$ is a prime, for every k, $m \leq k \leq n$.

Here is Euler's famous example: $f(X) = X^2 + X + 41$. For $k = 0$, 1, 2, 3, $\ldots$, 39, all its values are prime numbers, namely, 41, 43, 47, 53, 61, 71, 83, 97, 113, 131, 151, 173, 197, 223, 251, 281, 313, 347, 383, 421, 461, 503, 547, 593, 641, 691, 743, 797, 853, 911, 971, 1033,

1097, 1163, 1231, 1301, 1373, 1447, 1523, 1601. For $k = 40$ the value is $1681 = 41^2$.

For polynomials of the form $X^2 + X + q$, where q is a prime number, I note the interesting equivalent properties:

(1) $q = 2, 3, 5, 11, 17,$ or 41.

(2) $X^2 + X + q$ assumes prime values for $k = 0, 1, \ldots, q - 2$.

(3) The field $\mathbf{Q}(\sqrt{1 - 4q})$ of all algebraic numbers of the form $r + s\sqrt{1 - 4q}$ (where r, s are rational numbers) has "class number" 1 (which I will explain henceforth).

The implication (1) $\to$ (2) was stated by Euler, in 1772, in a letter to Bernoulli. Gauss proved, in *Disquisitiones Arithmeticae* (1801) that (1) implies (3). The equivalence of (2) and (3) was first proved by Rabinovitch in 1912. Heegner was the first to prove the much more difficult implication (3) $\to$ (1) (in 1952); however, his proof had some flaws. Later, Baker (1966), and Stark (1967) gave independent, flawless, and almost simultaneous proofs of the difficult implication. A detailed discussion of this result, omitting however the proof that (3) implies (1), may be found in Cohn's book (1962) or in my own article (1988).

And now, I elaborate on the meaning of condition (3). It is equivalent to say that each "algebraic integer" may be decomposed into "prime algebraic integers," in a unique way, except for "units." Thus, the arithmetic properties of integers in such fields are like those of ordinary integers, at least in respect to the fundamental theorem of factorization into primes. The main difference lies in the fact that the only integers that divide 1 are 1, -1, while in a field of algebraic numbers, there may be many more units, that is, algebraic integers dividing 1.

More involved results in the theory of quadratic number fields, which I must abstain from explaining, were used by Frobenius (1912) and Hendy (1974).

Here are some of their results:

The only polynomials $f(x) = 2X^2 + p$, with p prime, such that $f(k)$ is prime, for $k = 0, 1, 2, \ldots, p - 1$, are those where $p = 3, 5, 11, 29$. The only polynomials $f(x) = 2X^2 + 2X + \frac{p+1}{2}$, with p prime, $p \equiv 3 \pmod 4$, such that $f(k)$ is prime for $k = 0, 1, \ldots, p - 2$, are those for which $p = 3, 7, 19$.

These polynomials still do not surpass Euler's polynomial providing longer strings of successive prime values.

Record

For polynomials of the type $X^2 + X + q$ (with q prime) the best result is already Euler's, with $q = 41$, as follows from the preceding discussion. Calculations with quadratic polynomials $aX^2 + bX^2 + c$ have yielded several polynomials assuming 40 or more successive values at 0, 1, 2 .... As kindly communicated by R. Ruby, the records are: $(a, b, c) = (36, -810, 2753)$ giving 45 primes (discovered by R. Ruby); $(a, b, c) = (103, -3945, 34381)$, giving 43 prime values (discovered by R. Ruby); $(a, b, c) = (47, -1701, 10181)$ giving 43 prime values (discovered by G. Fung).

I shall return to polynomials of higher degree in Chapter 6.

For polynomials of degree one, $f(X) = dX + q$, there are at most q successive values $|f(0)|, |f(1)|, \ldots, |f(q-1)|$ which are primes. This leads, therefore, to the open problem:

Is it true that for every prime q there exists an integer $d \geq 1$ such that $q, d + q, 2d + q, \ldots, (q-1)d + q$ are primes? For example,

$q = 3, d = 2$ yield the primes 3, 5, 7;
$q = 5, d = 6$ yield the primes 5, 11, 17, 23, 29;
$q = 7, d = 150$ yield the primes 7, 157, 307, 457, 607, 757, 907.

This question is so difficult that I believe no one will prove it, nor find a counterexample in the near future.

Record

In 1986, G. Löh discovered that for $q = 11$ the smallest d is $d = 1536160080$, and for $q = 13$, the smallest d is $d = 9918821194590$.

I shall examine this and other questions about primes in arithmetic progressions in Chapter 4, Section IV.

Another investigation that has been undertaken is the search of quadratic polynomials $f(X)$ assuming prime values often. If $N > 1$, let $\nu(f, N) = \#\{ n \mid 0 \leq n \leq N$ such that $|f(n)|$ is equal to 1 or to a prime $\}$. The problem is to determine, for given N, the polynomial $f(X)$ for which $\nu(f, N)$ is maximum.

RECORD

For $N = 1000$ the polynomial $f(X) = 2X^2 - 199$, indicated by Karst in 1973, provides the maximum value of $\nu(f, 1000)$ up-to-date. Namely,

$$\nu(2X^2 - 199, 1000) = 598.$$

It should be noted that for Euler's polynomial

$$\nu(X^2 + X + 41, 1000) = 581.$$

I shall consider this question again in Chapter 6, Section IV.

In a race of cubic polynomials (like the 500 Miles of Indianapolis) masterminded by Goetgheluck (1989), the winner was:

$$f(X) = 2X^3 - 489X^2 + 39847X - 1084553,$$

which assumes 267 prime values for $n < 500$. Competitors have leading coefficient 1 or 2 and various size restrictions on the other coefficients.

Later, in Chapter 6, Section III, I shall also consider the contrary phenomenon of polynomials assuming composite values for all integers $n = 0, 1, 2, \ldots$ up to some large N.

III. Functions Satisfying Condition (c)

Surprisingly, if one requires only condition (c), the situation is totally different. This was discovered as a by-product in the investigation of Hilbert's tenth problem. The ideas come from logic and the results are quite extraord- inary—even if at the present they have not yet found immediate practical application.

In my presentation, I will not enter into technical details, which would take me too far from the prime numbers. Thus, I'll have to trade rigor for intuition, and I'm counting on the good will of the reader. Please do not interpret what I'm going to write in any undesirable way! The nice article of Davis (1973) is recommended for those intrigued with the results which will follow.

Hilbert's tenth problem asked about the solution in positive integers $(x_1, \ldots, x_n)$ of diophantine equations $P(X_1, \ldots, X_n) = 0$, where P is any polynomial with integral coefficients, and any number of indeterminates. More exactly: To give an algorithm that may be applied to any diophantine equation and will tell if it has solution in positive integers.

An algorithm should be understood as a definite procedure, which could be implemented as a computer program consisting of finitely many successive steps and leading to an answer "yes" or "no"—the kind of manipulations that mathematicians agree as legitimate. To the study of sets S of n-tuples $(x_1, \ldots, x_n)$ of positive integers, the central concept is the following: S is called a *diophantine set* if there exists a polynomial P with integral coefficients, in indeterminates $X_1, \ldots, X_n, Y_1, \ldots, Y_m$ ($m \geq 0$), such that $(x_1, \ldots, x_n) \in S$ if and only if there exist positive integers $y_1, \ldots, y_m$ satisfying

$$P(x_1, \ldots, x_n, y_1, \ldots, y_m) = 0.$$

First, the trivial examples. Every finite set S of n-tuples of positive integers is diophantine. Indeed, if S consists of the n-tuples $(a_1, a_2, \ldots, a_n)$, $(b_1, b_2, \ldots, b_n)$, $\ldots$, $(k_1, k_2, \ldots, k_n)$, let

$$
\begin{aligned}
P(X_1, X_2, \ldots, X_n) ={}& [(X_1 - a_1)^2 + (X_2 - a_2)^2 + \cdots + (X_n - a_n)^2] \\
&\times [(X_1 - b_1)^2 + (X_2 - b_2)^2 + \cdots + (X_n - b_n)^2] \\
&\vdots \\
&\times [(X_1 - k_1)^2 + (X_2 - k_2)^2 + \cdots + (X_n - k_n)^2].
\end{aligned}
$$

By its very definition, P has integral coefficients, so the set of all n-tuples $(x_1, \ldots, x_n)$ of positive integers such that $P(x_1, \ldots, x_n) = 0$, is a diophantine set.

Here is another example: the set S of all composite positive integers. Indeed, x is composite if and only if there exist positive integers y, z such that $x = (y+1)(z+1)$. Thus, x is composite whenever there exist y, z such that (x, y, z) is a solution of $X - (Y+1)(Z+1) = 0$.

The following fact, noted by Putnam in 1960, is not difficult to show:

A set S of positive integers is diophantine if and only if there exists a polynomial Q with integral coefficients (in $m \geq 1$ indeterminates) such that $S = \{ Q(x_1, \ldots, x_m) \geq 1 \mid x_1 \geq 1, \ldots, x_m \geq 1 \}$.

The next step in this theory consists in establishing that the set of prime numbers is diophantine.

For this purpose, it is necessary to examine the definition of prime numbers from the vantage of the theory of diophantine sets.

A positive integer x is a prime if and only if $x > 1$ and for any integers y, z such that $y \leq x$ and $z \leq x$, either $yz < x$, or $yz > x$, or $y = 1$, or $z = 1$. This definition of prime numbers contains bounded universally quantified occurrences of y, z, namely, $y \leq x$, $z \leq x$.

Another possible definition of prime numbers is the following. The positive integer x is a prime if and only if $x > 1$ and $\gcd((x-1)!, x) = 1$. The latter condition is rephrased as follows: There exist positive integers a, b such that $a(x - 1)! - bx = 1$; note that if a or b is negative, taking a sufficiently large integer k, then $a' = a + kx > 0$, $b' = b + k(x - 1)! > 0$ and $a'(x - 1)! - b'x = 1$.

Using one or the other characterization of prime numbers, it was shown with the theory developed by Putnam, Davis, J. Robinson and Matijasevič the important theorem:

The set of prime numbers is diophantine.

A combination of these results leads to the following astonishing result:

There exists a polynomial, with integral coefficients, such that the set of prime numbers coincides with the set of positive values taken by this polynomial, as the variables range in the set of nonnegative integers.

It should be noted that this polynomial also takes on negative values, and that a prime number may appear repeatedly as a value of the polynomial.

In 1971, Matijasevič indicated a system of algebraic relations leading to such a polynomial (without writing it explicitly) with degree 37, in 24 indeterminates; in the English translation of his paper this was improved to degree 21, and 21 indeterminates.

An explicit polynomial with this property, of degree 25, in the 26 indeterminates $a, b, c, \ldots, z$, was given by Jones, Sato, Wada & Wiens in 1976:

$$
\begin{aligned}
(k + 2)\{ &1 - [wz + h + j - q]^2 - [(gk + 2g + k + 1)(h + j) + h - z]^2 \\
&- [2n + p + q + z - e]^2 - [16(k + 1)^3(k + 2)(n - 1)^2 + 1 - f^2]^2 \\
&- [e^3(e + 2)(a + 1)^2 + 1 - o^2]^2 - [(a^2 - 1)y^2 + 1 - x^2]^2 \\
&- [16r^2y^4(a^2 - 1) + 1 - u^2]^2 - [((a + u^2(u^2 - a))^2 - 1)(n + 4dy)^2
\end{aligned}
$$

$$+ 1 - (x - cu)^2]^2 - [n + \ell + v - y]^2$$
$$- [(a^2 - 1)\ell^2 + 1 - m^2]^2 - [ai + k + 1 - \ell - i]^2$$
$$- [p + \ell(a - n - 1) + b(2an + 2a - n^2 - 2n - 2) - m]^2$$
$$- [q + y(a - p - 1) + s(2ap + 2a - p^2 - 2p - 2) - x]^2$$
$$- [z + p\ell(a - p) + t(2ap - p^2 - 1) - pm]^2 \}.$$

One is obviously tempted to reduce the number of indeterminates, or the degree, or both. But there is a price to pay. If the number n of indeterminates is reduced, then the degree d increases, and vice versa, if the degree d is forced to be smaller, the n must increase.

This is illustrated in the table concerning prime representing polynomials.

n = number of indeterminate	d = degree	Author	Year	Remarks
24	37	Matijasevič	1971	Not written explicitly
21	21	Same author	1971	
26	25	Jones, Sato, Wada & Wiens	1976	First explicit polynomial
42	5	Same authors	1976	Record low degree, not written explicitly
12	13697	Matijasevič	1976	
10	about 1.6×10^{45}	Same author	1977	Record low number of indeterminates, not written explicitly

It is not known which is the minimum possible number of variables (it cannot be 2). However, Jones showed that there is a prime representing polynomial of degree at most 5.

The same methods used to treat the set of prime numbers apply also to other diophantine sets once their defining arithmetical properties are considered from the appropriate point of view.

This has been worked out by Jones.

In a paper of 1975, Jones showed that the set of Fibonacci numbers is identical with the set of positive values at nonnegative integers, of the polynomial in 2 indeterminates and degree 5:

$$2xy^4 + x^2y^3 - 2x^3y^2 - y^5 - x^4y + 2y.$$

In 1979, Jones showed that each one of the sets of Mersenne primes, even perfect numbers, Fermat primes, corresponds in the same way to some polynomial in 7 indeterminates, however with higher degree. He also wrote explicitly other polynomials with lower degree, and more indeterminates, representing the above sets.

Set	Number of indeterminates	Degree
Fibonacci numbers	2	5
Mersenne primes	13	26
	7	914
Even perfect numbers	13	27
	7	915
Fermat primes	14	25
	7	905

By a method of Skolem (see his book, 1938), for the three latter sets the degree may be reduced to 5, however the number of variables increases to about 20.

For the set of Mersenne primes the polynomial is the following one, in the 13 indeterminates a, b, c, ... :

$$
\begin{aligned}
n\{\, 1 - &[4b + 3 - n]^2 - b([2 + hn^2 - a]^2 \\
&+ [n^3d^3(nd + 2)(h + 1)^2 + 1 - m^2]^2 \\
&+ [db + d + chn^2 + g(4a - 5) - kn]^2 + [(a^2 - 1)c^2 + 1 - k^2n^2]^2 \\
&+ [4(a^2 - 1)i^2c^4 + 1 - f^2]^2 \\
&+ [(kn + \ell f)^2 - ((a + f^2(f^2 - a))^2 - 1)(b + 1 + 2jc)^2 - 1]^2)\,\}.
\end{aligned}
$$

For the even perfect numbers, the polynomial in 13 indeterminates is

$$
\begin{aligned}
(2b + 2)n\{\, 1 - &[4b + 3 - n]^2 - b([2 + hn^2 - a]^2 \\
&+ n^3d^3(nd + 2)(h + 1)^2 + 1 - m^2]^2 \\
&+ [db + d + chn^2 + g(4a - 5) - kn]^2 \\
&+ [(a^2 - 1)c^2 + 1 - k^2n^2]^2 + [4(a^2 - 1)i^2c^4 + 1 - f^2]^2 \\
&+ [(kn + \ell f)^2 - ((a + f^2(f^2 - a))^2 - 1)(b + 1 + 2jc)^2 - 1]^2)\,\}.
\end{aligned}
$$

For the prime Fermat numbers, the polynomial in 14 indeterminates is

$$
\begin{aligned}
(6g + 5)\{\, 1 &- [bh + (a - 12)c + n(24a - 145) - d]^2 \\
&- [16b^3h^3(bh + 1)(a + 1)^2 + 1 - m^2]^2 \\
&- [3g + 2 - b]^2 - [2be + e - bh - 1]^2 - [k + b - c]^2 \\
&- [(a^2 - 1)c^2 + 1 - d^2]^2 - [4(a^2 - 1)i^2c^4 + 1 - f^2]^2 \\
&- [(d + \ell f)^2 - ((a + f^2(f^2 - a))^2 - 1)(b + 2jc)^2 - 1]^2 \,\}.
\end{aligned}
$$

4

How Are the Prime Numbers Distributed?

As I have already stressed, the various proofs of existence of infinitely many primes are not constructive and do not give an indication of how to determine the nth prime number. The proofs also do not indicate how many primes are less than any given number N. By the same token, there is no reasonable formula or function representing primes.

It will, however, be possible to predict with rather good accuracy the number of primes smaller than N (especially when N is large); on the other hand, the distribution of primes in short intervals shows a kind of built-in randomness. This combination of "randomness" and "predictability" yields at the same time an orderly arrangement and an element of surprise in the distribution of primes. According to Schroeder (1984), in his intriguing book *Number Theory in Science and Communication*, these are basic ingredients of works of art. Many mathematicians will readily agree that this topic has a great aesthetic appeal.

Recall from Chapter 3 that for every real number $x > 0$, $\pi(x)$ denotes the number of primes p such that $p \leq x$. $\pi(x)$ is also called the *prime counting function*.

The matters to consider are the following:

(I) The growth of $\pi(x)$, its order of magnitude, and comparison with other known functions.

(II) Results about the nth prime; the difference between consecutive primes, how small, how large, how irregular it may be. This includes the discussion of large gaps between consecutive primes, but leads also to several open problems, discussed below.

(III) Twin primes, their characterization and distribution.

(IV) Primes in arithmetic progressions.

 (V) Goldbach's famous conjecture.

(VI) The distribution of pseudoprimes and of Carmichael numbers.

Now I elaborate on these topics.

I. The Growth of $\pi(x)$

The basic idea in the study of the function $\pi(x)$, or others related to the distribution of primes, is to compare with functions that are both classical and computable, and such that their values are as close as possible to the values of $\pi(x)$. Of course, this is not simple and, as one might expect, an error will always be present. So, for each approximating function, one should estimate the order of magnitude of the difference, that is, of the error. The following notions are therefore natural.

Let $f(x)$, $h(x)$ be positive real valued continuous functions, defined for $x \geq x_0 > 0$.

The notation $f(x) \sim h(x)$ means that $\lim_{x \to \infty}(f(x)/h(x)) = 1$; $f(x)$ and $h(x)$ are then said to be asymptotically equal as x tends to infinity. Note that their difference may actually tend to infinity.

If, under the above hypothesis, there exist constants C, C', $0 < C < C'$, and x_0, x_1, with $x_1 \geq x_0$, such that $C \leq f(x)/h(x) \leq C'$ for all $x \geq x_1$, then $f(x)$, $h(x)$ are said to have the same order of magnitude.

If $f(x)$, $g(x)$, $h(x)$ are real valued continuous functions defined for $x \geq x_0 > 0$, and $h(x) > 0$ for all $x \geq x_0$, the notation

$$f(x) = g(x) + O(h(x))$$

means that the functions $f(x)$ and $g(x)$ have a difference that is ultimately bounded (as x tends to infinity) by a constant multiple of $h(x)$; that is, there exists $C > 0$ and $x_1 \geq x_0$ such that for every $x \geq x_1$ the inequality $|f(x) - g(x)| \leq Ch(x)$ holds. This is a useful notation to express the size of the error when $f(x)$ is replaced by $g(x)$.

Similarly, the notation

$$f(x) = g(x) + o(h(x))$$

means that $\lim_{x\to\infty}[f(x) - g(x)]/h(x) = 0$, so, intuitively, the error is negligible in comparison to $h(x)$.

A. History Unfolding

It is appropriate to describe in historical order the various discoveries about the distribution of primes, culminating with the prime number theorem. This is how Landau proceeded in his famous treatise *Handbuch der Lehre von der Verteilung der Primzahlen*, which is the classical work on the subject.

Euler

First, I give a result of Euler which tells, not only that there are infinitely many primes, but also that "the primes are not so sparse as the squares." (This statement will be made clear shortly.)

Euler noted that for every real number $\sigma > 1$ the series $\sum_{n=1}^{\infty}(1/n^{\sigma})$ is convergent, and in fact, for every $\sigma_0 > 1$ it is uniformly convergent on the half-line $\sigma_0 \le x < \infty$. Thus, it defines a function $\zeta(\sigma)$ (for $1 < \sigma < \infty$), which is continuous and differentiable. Moreover, $\lim_{\sigma\to\infty} \zeta(\sigma) = 1$ and $\lim_{\sigma\to 1+0}(\sigma - 1)\zeta(\sigma) = 1$. The function $\zeta(\sigma)$ is called the *zeta function*.

The link between the zeta function and the prime numbers is the following Eulerian product, which expresses the unique factorization of integers as product of primes:

$$\sum_{n=1}^{\infty} \frac{1}{n^{\sigma}} = \prod_{p} \frac{1}{1 - \frac{1}{p^{\sigma}}} \qquad \text{(for } \sigma > 1\text{)}.$$

In particular, this implies that $\zeta(\sigma) \ne 0$ for $\sigma > 1$.

With the same idea used in his proof of the existence of infinitely many primes (see Chapter 1), Euler proved in 1737:

The sum of the inverses of the prime numbers is divergent:

$$\sum_{p} \frac{1}{p} = \infty.$$

Proof. Let N be an arbitrary natural number. Each integer $n \le N$ is a product, in a unique way, of powers of primes p, $p \le n \le N$.

Also for every prime p,

$$\sum_{k=1}^{\infty} \frac{1}{p^k} = \frac{1}{1 - \frac{1}{p}}.$$

Hence

$$\sum_{n=1}^{N} \frac{1}{n} \leq \prod_{p \leq N} \left(\sum_{k=1}^{\infty} \frac{1}{p^k} \right) = \prod_{p \leq N} \frac{1}{1 - \frac{1}{p}}.$$

But

$$\log \prod_{p \leq N} \frac{1}{1 - \frac{1}{p}} = -\sum_{p \leq N} \log \left(1 - \frac{1}{p} \right),$$

and for each prime p,

$$- \log \left(1 - \frac{1}{p} \right) = \sum_{m=1}^{\infty} \frac{1}{m p^m} \leq \frac{1}{p} + \frac{1}{p^2} \left(\sum_{h=0}^{\infty} \frac{1}{p^h} \right)$$

$$= \frac{1}{p} + \frac{1}{p^2} \times \frac{1}{1 - \frac{1}{p}} = \frac{1}{p} + \frac{1}{p(p-1)} < \frac{1}{p} + \frac{1}{p^2}.$$

Hence

$$\log \sum_{n=1}^{N} \frac{1}{n} \leq \log \prod_{p \leq N} \frac{1}{1 - \frac{1}{p}} \leq \sum_{p \leq N} \frac{1}{p} + \sum_{p \leq N} \frac{1}{p^2} \leq \sum_{p} \frac{1}{p} + \sum_{n=1}^{\infty} \frac{1}{n^2}.$$

But the series $\sum_{n=1}^{\infty} (1/n^2)$ is convergent. Since N is arbitrary and the harmonic series is divergent, then $\log \sum_{n=1}^{\infty} (1/n) = \infty$, and therefore the series $\sum_p (1/p)$ is divergent. ☐

As I have already mentioned, the series $\sum_{n=1}^{\infty} (1/n^2)$ is convergent. Thus, it may be said, somewhat vaguely, that the primes are not so sparsely distributed as the squares.

One of the beautiful discoveries of Euler was the sum of this series:

$$\sum_{n=1}^{\infty} \frac{1}{n^2} = \frac{\pi^2}{6}.$$

Euler also evaluated the sums $\sum_{n=1}^{\infty} (1/n^{2k})$ for every $k \geq 1$, thereby solving a rather elusive problem.

For this purpose, he made use of the Bernoulli numbers, which are defined as follows:

$$B_0 = 1, \quad B_1 = -\frac{1}{2}, \quad B_2 = \frac{1}{6}, \dots,$$

B_k being recursively defined by the relation

$$\binom{k+1}{1}B_k + \binom{k+1}{2}B_{k-1} + \cdots + \binom{k+1}{k}B_1 + B_0 = 0.$$

These numbers are clearly rational, and it is easy to see that $B_{2k+1} = 0$ for every $k \geq 1$. They appear also as coefficients in the Taylor expansion:

$$\frac{x}{e^x - 1} = \sum_{k=0}^{\infty} \frac{B_k}{k!}x^k.$$

Using Stirling's formula,

$$n! \sim \frac{\sqrt{2\pi}n^{n+\frac{1}{2}}}{e^n} \qquad (\text{as } n \to \infty),$$

it may also be shown that

$$|B_{2n}| \sim 4\sqrt{\pi n}\left(\frac{n}{\pi e}\right)^{2n},$$

hence the above series is convergent in the interval $|x| < 2\pi$.

Euler had already used the Bernoulli numbers to express the sums of equal powers of consecutive numbers:

$$\sum_{j=1}^{n} j^k = S_k(n)(k \geq 1),$$

where

$$S_k(X) = \frac{1}{k+1}\left[X^{k+1} - \binom{k+1}{1}B_1 X^k\right.$$

$$\left. + \binom{k+1}{2}B_2 X^{k-1} + \cdots + \binom{k+1}{k}B_k X\right].$$

A similar expression was also obtained by Seki in Japan at about the same time.

Euler's formula giving the value of $\zeta(2k)$ is:

$$\zeta(2k) = \sum_{n=1}^{\infty} \frac{1}{n^{2k}} = (-1)^{k+1}\frac{(2\pi)^{2k}B_{2k}}{2(2k)!}.$$

In particular,

$$\zeta(2) \;=\; \sum_{n=1}^{\infty} \frac{1}{n^2} = \frac{\pi^2}{6} \qquad \text{(already mentioned)},$$

$$\zeta(4) \;=\; \sum_{n=1}^{\infty} \frac{1}{n^4} = \frac{\pi^4}{90}, \text{ etc.}$$

Euler also considered the Bernoulli polynomials, defined by

$$B_k(X) = \sum_{i=0}^{k} \binom{k}{i} B_i X^{k-i} \qquad (k \geq 0).$$

They may be used to rewrite the expression for $S_k(X)$, but more important is their application to a far reaching generalization of Abel's summation formula, namely, the well-known Euler-MacLaurin summation formulas:

If $f(x)$ is a continuous function, continuously differentiable as many times as required, if $a < b$ are integers, then

$$\sum_{n=a+1}^{b} f(n) \;=\; \int_a^b f(t)dt + \sum_{r=1}^{k}(-1)^r \frac{B_r}{r!}\{ f^{(r-1)}(b) - f^{(r-1)}(a) \}$$

$$+ \frac{(-1)^{k-1}}{k!} \int_a^b B_k(t - [t]) f^{(k)}(t)dt.$$

The reader is urged to consult the paper by Ayoub, *Euler and the zeta function* (1974), where there is a description of the many imaginative relations and findings of Euler concerning $\zeta(s)$—some fully justified, others only made plausible, but anticipating later works by Riemann.

LEGENDRE

The first serious attempt to study the function $\pi(x)$ is due to Legendre (1808), who used the Eratosthenes sieve and proved that

$$\pi(N) = \pi(\sqrt{N}) - 1 + \sum \mu(d) \left[\frac{N}{d}\right].$$

The notation $[t]$ has already been explained, the summation is over all divisors d of the product of all primes $p \leq \sqrt{N}$, and $\mu(n)$

denotes the Möbius function, which was already defined in Chapter 3, Section I.

As a consequence, Legendre showed that $\lim_{x\to\infty}(\pi(x)/x) = 0$, but this is a rather weak result.

Experimentally, Legendre conjectured in 1798 and again in 1808 that

$$\pi(x) = \frac{x}{\log x - A(x)},$$

where $\lim_{x\to\infty} A(x) = 1.08366\ldots$. That Legendre's statement cannot be true, was shown forty years later by Tschebycheff (see below). An easy proof was given by Pintz (1980).

GAUSS

At age 15, in 1792, Gauss conjectured that $\pi(x)$ and the function logarithmic integral of x, defined by

$$\text{Li}(x) = \int_2^x \frac{dt}{\log t},$$

are asymptotically equal. Since $\text{Li}(x) \sim x/\log x)$, this may be written also as

$$\pi(x) \sim \frac{x}{\log x}.$$

This conjecture was to be confirmed later, and is now known as the prime number theorem; I shall soon return to this matter.

The approximation of $\pi(x)$ by $x/\log x$ is only reasonably good, while it is much better using the logarithmic integral, as it will be illustrated in a table.

TSCHEBYCHEFF

Important progress for the determination of the order of magnitude of $\pi(x)$ was made by Tschebycheff, around 1850. He proved, using elementary methods, that there exist constants C, C', $0 < C' < 1 < C$, such that

$$C'\frac{x}{\log x} < \pi(x) < C\frac{x}{\log x} \qquad \text{(for } x \geq 2\text{)}.$$

He actually computed values for C, C' very close to 1. For example, taking $x \geq 30$,

$$C' = \log \frac{2^{1/2}3^{1/3}5^{1/5}}{30^{1/30}} = 0.92129\ldots,$$

$C = \frac{6}{5}C' = 1.10555\ldots$. Moreover, if the limit of

$$\frac{\pi(x)}{x/\log x}$$

exists (as $x \to \infty$), it must be equal to 1. He deduced also that Legendre's approximation of $\pi(x)$ cannot be true, unless 1.08366 is replaced by 1 (see Landau's book, page 17).

Tschebycheff also proved Bertrand's postulate that between any natural number $n \geq 2$ and its double there exists at least one prime. I shall discuss this proposition in more detail, when I present the main properties of the function $\pi(x)$.

Tschebycheff worked with the function $\theta(x) = \sum_{p \leq x} \log p$, now called *Tschebycheff's function*, which yields basically the same information as $\pi(x)$, but is somewhat easier to handle.

Even though Tschebycheff came rather close, the proof of the fundamental prime number theorem, conjectured by Gauss, had to wait for about 50 more years, until the end of the century. During this time, important new ideas were contributed by Riemann.

RIEMANN

Riemann had the idea of defining the zeta function for complex numbers s having real parts greater than 1, namely,

$$\zeta(s) = \sum_{n=1}^{\infty} \frac{1}{n^s}$$

The Euler product formula still holds, for every complex s with $\mathrm{Re}(s) > 1$.

Using the Euler-MacLaurin summation formula, $\zeta(s)$ is expressible as follows:

$$\zeta(s) = \frac{1}{s-1} + \frac{1}{2} + \sum_{r=2}^{k} \frac{B_r}{r!} s(s+1)\cdots(s+r-2)$$
$$- \frac{1}{k!} s(s+1)\cdots(s+k-1) \int_{1}^{\infty} B_k(x-[x]) \frac{dx}{x^{s+k}}.$$

Here k is any integer, $k \geq 1$, the numbers B_r are the Bernoulli numbers, which the reader should not confuse with $B_k(x - [x])$, the value of the kth Bernoulli polynomial $B_k(X)$ at $x - [x]$.

The integral converges to $\mathrm{Re}(s) > 1-k$, and since k is an arbitrary natural number, this formula provides the analytic continuation of $\zeta(s)$ to the whole plane.

$\zeta(s)$ is everywhere holomorphic, except at $s = 1$, where it has a simple pole with residue 1, that is, $\lim_{s \to 1}(s - 1)\zeta(s) = 1$.

In 1859, Riemann established the functional equation for the zeta function. Since this equation involves the gamma function $\Gamma(s)$, I must first define $\Gamma(s)$. For $\mathrm{Re}(s) > 0$, a convenient definition is by means of the Eulerian integral

$$\Gamma(s) = \int_0^\infty e^{-u} u^{s-1} du.$$

For arbitrary complex numbers s, it may be defined by

$$\Gamma(s) = \frac{1}{se^{\gamma s}} \prod_{n=1}^{\infty} \frac{e^{s/n}}{1 + \frac{s}{n}},$$

where γ is Euler's constant, equal to

$$\gamma = \lim_{n \to \infty} \left(1 + \frac{1}{2} + \cdots + \frac{1}{n} - \log n \right) = 0.577215665\ldots.$$

Euler's constant, also known with good reason as Mascheroni's constant, by the Italians, is related to Euler's product by the following formula of Mertens:

$$e^\gamma = \lim_{n \to \infty} \frac{1}{\log n} \prod_{i=1}^{n} \frac{1}{1 - \frac{1}{p_i}}.$$

$\Gamma(s)$ is never equal to 0; it is holomorphic everywhere except at the points $0, -1, -2, -3, \ldots$, where it has simple poles. For every positive integer n, $\Gamma(n) = (n-1)!$, so the gamma function is an extension of the factorial function. The gamma function satisfies many interesting relations, among which are the functional equations

$$\Gamma(s)\Gamma(1-s) = \frac{\pi}{\sin \pi s} \quad \text{and} \quad \Gamma(s)\Gamma\left(s + \frac{1}{2}\right) = \frac{\sqrt{\pi}}{2^{2s-1}}\Gamma(2s),$$
$$\Gamma(s+1) = s\Gamma(s).$$

Here is the functional equation for the Riemann zeta function:

$$\pi^{-s/2}\Gamma\left(\frac{s}{2}\right)\zeta(s) = \pi^{-(1-s)/2}\Gamma\left(\frac{1-s}{2}\right)\zeta(1-s).$$

For example, it follows from the functional equation that $\zeta(0) = -\frac{1}{2}$.

The zeros of the zeta function are:

(a) Simple zeroes at the points $-2, -4, -6, \ldots$, which are called the trivial zeroes.

(b) Zeroes in the critical strip, consisting of the nonreal complex numbers s with $0 \leq \operatorname{Re}(s) \leq 1$.

Indeed, if $\operatorname{Re}(s) > 1$, then by the Euler product, $\zeta(s) \neq 0$. If $\operatorname{Re}(s) < 0$, then $\operatorname{Re}(1 - s) > 1$, the right-hand side in the functional equation is not zero, so the zeroes must be exactly at $s = -2, -4, -6, \ldots$, which are the poles of $\Gamma(s/2)$.

The knowledge of the zeroes in the critical strip has a profound influence on the understanding of the distribution of primes. A first thing to note is that the zeroes in the critical strip are not real and they are symmetric about the real axis and the vertical line $\operatorname{Re}(x) = \frac{1}{2}$.

Riemann conjectured that all nontrivial zeroes ρ of $\zeta(s)$ are on the critical line $\operatorname{Re}(s) = \frac{1}{2}$, that is, $\rho = \frac{1}{2} + i\gamma$. This is the famous *Riemann's hypothesis*, which has never been proved. It is undoubtedly a very difficult and important problem.

Riemann also had the idea of considering all the powers of primes $p^n \leq x$, with each such p weighted as $1/n$. For this purpose, he defined the function

$$
J(x) = \begin{cases}
\pi(x) + \frac{1}{2}\pi(x^{1/2}) + \frac{1}{3}\pi(x^{1/3}) + \frac{1}{4}\pi(x^{1/4}) + \cdots - \\
\quad \frac{1}{2m}, \text{ if } x = p^m, \text{ where } p \text{ is a prime number,} \\
\quad m \geq 1. \\
\pi(x) + \frac{1}{2}\pi(x^{1/2}) + \frac{1}{3}\pi(x^{1/3}) + \frac{1}{4}\pi(x^{1/4}) + \cdots, \\
\quad \text{if } x > 0 \text{ is a real number which is neither a} \\
\quad \text{prime nor a prime-power.}
\end{cases}
$$

One of the principal formulas conjectured by Riemann was an expression of $J(x)$ in terms of the logarithmic integral; this formula involves the zeroes of $\zeta(s)$.

First, define $\operatorname{Li}(e^w)$ for any complex number $w = u + iv$, as follows.

$$
\operatorname{Li}(e^w) = \int \frac{e^t}{t} dt + z,
$$

where the integral is over the horizontal line, from $-\infty$ to $u + iv$, and $z = \pi i$, $-\pi i$, 0, according to $v > 0$, $v < 0$, or $v = 0$. Riemann's formula, which was proved by von Mangoldt, is

$$J(x) = \text{Li}(x) - \sum \text{Li}(x^\rho) + \int_x^\infty \frac{dt}{t(t^2 - 1)\log t} \log 2$$

[the sum is extended over all the nontrivial zeroes ρ of $\zeta(s)$, each with its own multiplicity].

Let

$$R(x) = \sum_{m=1}^\infty \frac{\mu(m)}{m} \text{Li}(x^{1/m})$$

be the so-called Riemann function.

Riemann gave the following exact formula for $\pi(x)$, in terms of the Riemann function:

$$\pi(x) = R(x) - \sum_\rho R(x^\rho)$$

[the sum being extended over all the nontrivial zeroes of $\zeta(x)$, each counted with its own multiplicity].

The Riemann function $R(x)$ provides a very good approximation for $\pi(x)$, as will be seen in the following table. The size of the error is expressed in terms of $R(x^\rho)$, for all the roots of $\zeta(s)$ in the critical strip.

The Riemann function is computable by this quickly converging power series, given by Gram in 1893:

$$R(x) = 1 + \sum_{n=1}^\infty \frac{1}{n\zeta(n+1)} \times \frac{(\log x)^n}{n!}.$$

The work of Riemann on the distribution of primes is thoroughly studied in Edwards' book (1974), which I recommend without reservations. Other books on the Riemann zeta function are the classical treatise by Titchmarsh (1951) and the recent volumes of Ivić (1985) and Patterson (1988).

DE LA VALLÉE POUSSIN AND HADAMARD

Riemann provided many of the tools for the proof of the fundamental *prime number theorem*:

$$\pi(x) \sim \frac{x}{\log x}.$$

Other tools came from the theory of complex analytic functions, which was experiencing a period of rapid growth.

The prime number theorem was raised to the status "a most wanted theorem," and it was folklore to consider that he who would prove it would become immortal.

The theorem was established, not by one, but by two eminent analysts, independently, and in the same year (1896). No, they did not become immortals, as in some old Greek legend, but ... almost! Hadamard lived to the age of 98, de la Vallée Poussin just slightly less, to the age of 96.

de la Vallée Poussin established the following fact: there exists $c > 0$ and $t_0 = t_0(c) > e^{2c}$, such that $\zeta(s) \neq 0$ for every $s = \sigma + it$ in the region:

$$\begin{cases} 1 - \frac{c}{\log t_0} \leq \sigma \leq 1, & \text{when } |t| \leq t_0 \\ \\ 1 - \frac{c}{\log |t|} \leq \sigma \leq 1, & \text{when } t_0 \leq |t|. \end{cases}$$

Thus, in particular, $\zeta(1+it) \neq 0$ for every t, as shown by Hadamard. The determination of a large zero-free region for $\zeta(s)$ was an important feature in the proof of the prime number theorem.

Not only did Hadamard and de la Vallée Poussin prove the prime number theorem. They have also estimated the error as being:

$$\pi(x) = \text{Li}(x) + O(xe^{-A\sqrt{\log x}}),$$

for some positive constant A.

I shall soon tell how the error term was subsequently reduced by determining larger zero-free regions for the zeta function.

There have been many proofs of the prime number theorem with analytical methods. One which is particularly simple is due to Newman (1980).

There are other equivalent ways of formulating the prime number theorem. Using the Tschebycheff function, the theorem may be rephrased as follows:

$$\theta(x) \sim x.$$

Another formulation involves the summatory function of the *von Mangoldt function*. Let

$$\Lambda(n) = \begin{cases} \log p & \text{if } n = p^\nu \ (\nu \geq 1) \text{ and } p \text{ is a prime} \\ 0 & \text{otherwise.} \end{cases}$$

This function, appears in the expression of the logarithmic derivative of the zeta function:

$$-\frac{\zeta'(s)}{\zeta(s)} = \sum_{n=1}^{\infty} \frac{\Lambda(n)}{n^s} \qquad [\text{for } \mathrm{Re}(s) > 1].$$

It is also related to the function $J(x)$ already encountered:

$$J(x) = \sum_{n \le x} \frac{\Lambda(n)}{\log n}.$$

The summatory function of $\Lambda(n)$ is defined to be

$$\psi(x) = \sum_{n \le x} \Lambda(n).$$

It is easily expressible in terms of Tschebycheff's function

$$\psi(x) = \theta(x) + \theta(x^{1/2}) + \theta(x^{1/3}) + \cdots.$$

The prime number theorem may also be formulated as:

$$\psi(x) \sim x.$$

ERDÖS AND SELBERG

It was believed for a long time that analytical methods could not be avoided in the proof of the prime number theorem. Thus, the mathematical community was surprised when both Erdös and Selberg showed, in 1949, how to prove the prime number theorem using essentially only elementary estimates of arithmetical functions.

Many such estimates of sums were already known, as, for example,

$$\sum_{n \le x} \frac{1}{n} = \log x + \gamma + O\left(\frac{1}{x}\right), \qquad \text{where } \gamma \text{ is Euler's constant;}$$

$$\sum_{n \le x} \frac{1}{n^\sigma} = \frac{x^{1-\sigma}}{1-\sigma} + \zeta(\sigma) + O\left(\frac{1}{x^\sigma}\right), \qquad \text{where } \sigma > 1;$$

$$\sum_{n \le x} \log n = x \log x - x + O(\log x);$$

$$\sum_{n \le x} \frac{\log n}{n} = \frac{1}{2}(\log x)^2 + C + O\left(\frac{\log x}{x}\right).$$

The above estimates are obtained using the Abel or Euler–Maclaurin summation formulas, and have really no arithmetical content. The following sums involving primes are more interesting:

$$\sum_{p\leq x} \frac{\log p}{p} = \log x + O(1);$$

$$\sum_{p\leq x} \frac{1}{p} = \log\log x + C + O\left(\frac{1}{\log x}\right); \text{where } C = 0.2615\ldots;$$

$$\sum_{n\leq x} \frac{\Lambda(n)}{n} = \log x + O(1);$$

$$\sum_{n\leq x} \frac{\Lambda(n)\log n}{n} = \frac{1}{2}(\log x)^2 + O(\log x).$$

Selberg gave in 1949 the following estimate:

$$\sum_{p\leq x}(\log p)^2 + \sum_{pq\leq x}(\log p)(\log q) = 2x\log x + O(x)$$

(where p, q are primes).

This estimate is, in fact, equivalent to each of the following:

$$\theta(x)\log x + \sum_{p\leq x}\theta\left(\frac{x}{p}\right)\log p = 2x\log x + O(x);$$

$$\sum_{n\leq x}\Lambda(n)\log n + \sum_{mn\leq x}\Lambda(m)\Lambda(n) = 2x\log x + O(x).$$

From his estimate, Selberg was able to give an elementary proof of the prime number theorem. At the same time, also using a variant of Selberg's estimate

$$\frac{\psi(x)}{x} + \frac{1}{\log x}\sum_{n\leq x}\frac{\psi(x/n)\Lambda(n)}{x/n}\frac{\Lambda(n)}{n} = 2 + O\left(\frac{1}{\log x}\right),$$

Erdös gave, with a different elementary method, his proof of the prime number theorem.

B. Sums Involving the Möbius Function

Even before Möbius had formally defined the function $\mu(n)$, Euler had already considered it. In 1748, based on experimental evidence, Euler conjectured that $\sum_{n=1}^{\infty}\frac{\mu(n)}{n}$ converges to 0. von Mangoldt proved this conjecture, as an application of the prime number theorem. Also for every s with $Re(s) > 1$,

$$\sum_{n=1}^{\infty} \frac{\mu(n)}{n^s} = \frac{1}{\zeta(s)}.$$

Actually, the converse is also true, so the prime number theorem is equivalent to $\sum_{n=1}^{\infty} \frac{\mu(n)}{n} = 0$.

It follows that for every $x > 1$,

$$\sum_{n \leq x} \frac{\mu(n)}{n^2} = \frac{6}{\pi^2} + O\left(\frac{1}{x}\right).$$

The summatory function of the Möbius function is the *Mertens function*

$$M(x) = \sum_{n \leq x} \mu(n).$$

It may be shown that the prime number theorem is also equivalent to the assertion that $\lim_{x \to \infty} M(x)/x = 0$. For details relative to the preceding statements, the reader may consult the books of Landau (1909) Ayoub (1963) or Apostol (1976).

C. The Distribution of Values of Euler's Function

I will gather here results concerning the distribution of values of Euler's function. They supplement the properties already stated in Chapter 2, Section II.

First, some indications about the growth of Euler's function. It is easy to show that

$$\phi(n) \geq \log 2 \frac{n}{\log(2n)},$$

in particular, for every $\delta > 0$, $\phi(n)$ grows ultimately faster than $n^{1-\delta}$.

Even better, for every $\varepsilon > 0$ there exists $n_0 = n_0(\varepsilon)$ such that, if $n \geq n_0$, then

$$\phi(n) \geq (1 - \varepsilon)e^{-\gamma} \frac{n}{\log \log n}.$$

On the other hand, it follows from the prime number theorem, that there exist infinitely many n such that

$$\phi(n) < (1 + \varepsilon)e^{-\gamma} \frac{n}{\log \log n}.$$

So,

$$\lim\inf \frac{\phi(n)\log\log n}{n} = e^{-\gamma}.$$

A proof of the above results may be found, for example, in the books by Landau (1909), or Apostol (1976).

What is the average of $\phi(n)$?

From the relation $n = \sum_{d|n}\phi(d)$, it is not difficult to show that

$$\frac{1}{x}\sum_{n\le x}\phi(n) = \frac{3x}{\pi^2} + O(\log x).$$

So, the mean value of $\phi(n)$ is equal to $3n/\pi^2$.

As a consequence, if two integers m, $n \ge 1$ are randomly chosen, then the probability that m, n be relatively prime is $6/\pi^2$.

All these matters are well explained in the books of Hardy & Wright and Apostol (1976).

D. TABLES OF PRIMES

Now, I turn my attention to tables of prime numbers, and of factors of numbers (not divisible by 2, 3, or 5). The first somewhat extended tables are by Brancker in 1668 (table of least factor of numbers up to 100,000), Krüger in 1746 (primes up to 100,000), Lambert in 1770 (table of least factor of numbers up to 102,000), Felkel in 1776 (table of least factor of numbers up to 408,000), Vega in 1797 (primes up to 400,031), Chernac in 1811 (prime factors of numbers up to 1,020,000), and Burkhardt in 1816/7 (least factor of numbers up to 3,036,000).

Legendre and Gauss based their empirical observations on the available tables.

Little by little, the tables were extended. Thus in 1856 Crelle presented to the Berlin Academy a table of primes up to 6,000,000, and this work was extended by Dase, before 1861, up to 9,000,000, entirely done by mental calculation!

But in this connection, the most amazing feat is Kulik's factor table of numbers to 100,330,200 (except for multiples of 2, 3, 5), entitled *Magnus Canon Divisorum pro omnibus numeris per 2, 3 et 5 non divisibilibus, et numerorum primorum interfacentium ad millies centena millia accuratius ad 100330201 usque.* Kulik spent about 20 years preparing this table, and at his death in 1863, the eight manuscript volumes, with a total of 4212 pages, were deposited at the Academy of Sciences in Vienna (in February 1867).

In 1909, D. N. Lehmer published a table of factor numbers up to about 10,000,000, and in 1914 he published the list of primes up to that limit. This time, the volumes were widely distributed and easily accessible to mathematicians.

With the advent of computers, numerous tables were prepared and published; some were available on cards or tapes.

It should be noted that tables of primes on cards or tape are obsolete, because it is possible to generate all the primes up to any given bound, or in a prescribed interval, with the sieve of Eratosthenes, quicker than one can read it from card or tape.

To my readers, who have faithfully arrived up to this point, as a token of appreciation, and for their utmost convenience, I include a TABLE OF PRIMES UP TO 10000 following the Bibliography! Enjoy yourself!

E. THE EXACT VALUE OF $\pi(x)$ AND COMPARISON WITH $x/(\log x)$, $\mathrm{Li}(x)$, AND $R(x)$

The exact values of $\pi(x)$ may be obtained by direct counting using tables, or by an ingenious "telescoping" method devised in 1871 by Meissel, a German astronomer, which allowed him to go far beyond the range of the tables. In fact, to compute $\pi(x)$ the method requires the knowledge of the prime numbers $p \leq x^{1/2}$ as well as the values of $\pi(y)$ for $y \leq x^{2/3}$. It is based on the following formula

$$\pi(x) = \phi(x, m) + m(s + 1) + \frac{s(s-1)}{2} - 1 - \sum_{i=1}^{s} \pi\left(\frac{x}{p_{m+i}}\right),$$

where $m = \pi(x^{1/3})$, $n = \pi(x^{1/2})$, $s = n - m$, and $\phi(x, m)$ denotes the number of integers a such that $a \leq x$ and a is not a multiple of $2, 3, \ldots, p_m$.

Even though the calculation of $\phi(x, m)$ is long, when m is large, it offers no major difficulty. The calculation is based on the following simple facts:

Recurrence relation:

$$\phi(x, m) = \phi(x, m - 1) - \phi\left(\left[\frac{x}{p_m}\right], m - 1\right).$$

Division property:

If $P_m = p_1 p_2 \cdots p_m$, if $a \geq 0$, $0 \leq r < P_m$, then $\phi(aP_m + r, m) = a\phi(P_m) + \phi(r, m)$.

Symmetry property:

If $\frac{1}{2}P_m < r < P_m$, then

$$\phi(r, m) = \phi(P_m) - \phi(P_m - r - 1, m).$$

Meissel determined, in 1885, the number $\pi(10^9)$ (however he found a value which is low by 56). A simple proof of Meissel's formula was given by Brauer in 1946.

RECORD

The largest computed value of $\pi(x)$ is $\pi(4 \times 10^{16}) =$ 1,075,292,778,753,150, by Lagarias, Miller & Odlyzko (1985).
  The following table gives values of $\pi(x)$, which may be compared with the calculated values of the functions $x/(\log x)$, $\mathrm{Li}(x)$, $R(x)$.

x	$\pi(x)$	$\left[\frac{x}{\log x}\right] - \pi(x)$	$[\mathrm{Li}(x)] - \pi(x)$	$[R(x)] - \pi(x)$
10^8	5,761,455	$-332,774$	754	97
10^9	50,847,534	$-2,592,592$	1,701	-79
10^{10}	455,052,511	$-20,758,030$	3,104	$-1,828$
10^{11}	4,118,054,813	$-169,923,160$	11,588	$-2,318$
10^{12}	37,607,912,018	$-1,416,706,193$	38,263	$-1,476$
10^{13}	346,065,536,839	$-11,992,858,452$	108,971	$-5,773$
10^{14}	3,204,941,750,802	$-102,838,308,636$	314,890	$-19,200$
10^{15}	29,844,570,422,669	$-891,604,962,453$	1,052,619	73,218
10^{16}	279,238,341,033,925	$-7,804,289,844,393$	3,214,632	327,052
2×10^{16}	547,863,431,950,008	$-15,020,437,343,198$	3,776,488	$-225,875$
4×10^{16}	1,075,292,778,753,150	$-28,929,900,579,950$	5,538,861	$-10,980$

I have already mentioned Tschebycheff's inequalities for $\pi(x)$, obtained with elementary methods and prior to the prime number theorem. In 1892, Sylvester refined Tschebycheff's method, obtaining

$$0.95695 \frac{x}{\log x} < \pi(x) < 1.04423 \frac{x}{\log x}$$

for every x sufficiently large (see also Langevin, 1977). For teaching purposes, there is a very elegant determination by Erdös (in 1930) of the appropriate constants; thus

$$\log 2 \frac{x}{\log x} < \pi(x) < 2 \log 2 \frac{x}{\log x}$$

for every sufficiently large x.

In 1962, using a very delicate analysis, Rosser & Schoenfeld showed that, if $x \geq 17$, then $x/(\log x) \leq \pi(x)$.

On the other hand, Riemann and Gauss believed that $\text{Li}(x) > \pi(x)$ for every sufficiently large x. Even though in the present range of tables this is true, it had been shown by Littlewood in 1914 that the difference $\text{Li}(x) - \pi(x)$ changes sign infinitely often, say, at numbers $x_0 < x_1 < x_2 < \cdots$, where x_n tends to infinity.

Assuming Riemann's hypothesis, Skewes showed in 1933 that $x_0 < 10^{10^{10^{34}}}$. For a long time, this number was famous, as being the largest number that appeared in a somewhat natural way in mathematics. One knows now, even without assuming Riemann's hypothesis, a much smaller upper bound for x_0.

RECORD

In a computation reported in 1986, te Riele has showed that already between 6.62×10^{370} and 6.69×10^{370} there are more than 10^{180} successive integers for which $\text{Li}(x) < \pi(x)$.

F. THE NONTRIVIAL ZEROES OF $\zeta(s)$

I recall that the zeros of the Riemann zeta function are the trivial zeroes $-2, -4, -6, \ldots$ and the nontrivial zeroes $\sigma + it$, with $0 \leq \sigma \leq 1$, that is, zeroes in the critical strip.

First, I shall discuss the zeroes in the whole critical strip and then the zeroes on the critical line $\text{Re}(s) = \frac{1}{2}$.

Since $\zeta(\bar{s}) = \overline{\zeta(s)}$ (where the bar denotes the complex conjugate), then the zeroes lie symmetrically with respect to the real axis; so, it suffices to consider the zeroes in the upper half of the critical strip.

Since there can only be finitely many zeroes $\sigma + it$ for each value $t > 0$, the zeroes of $\zeta(s)$ in the upper half of the critical strip may be enumerated as $\rho_n = \sigma_n + it_n$ with $0 < t_1 \leq t_2 \leq t_3 \leq \cdots$.

For every $T > 0$, let $N(T)$ denote the number of zeroes $\rho_n = \sigma + n + it_n$ in the critical strip, with $0 < t_n \leq T$.

Similarly, let $N_0(T)$ denote the number of zeroes $\frac{1}{2} + it$ of Riemann's zeta function, which lie on the critical line, such that $0 < t \leq T$.

Clearly, $N_0(T) \leq N(T)$ and Riemann's hypothesis is the statement that $N_0(T) = N(T)$ for every $T > 0$.

Here are the main results concerning $N(T)$. First of all, it was conjectured by Riemann, and proved by von Mangoldt:

$$N(T) = \frac{T}{2\pi} \left\{ \log \left[\frac{T}{2\pi} \right] - 1 \right\} + 0(\log T).$$

It follows that there exist infinitely many zeroes in the critical strip.

All the known nontrivial zeroes of $\zeta(s)$ are simple and lie on the critical line. Montgomery showed in 1973, assuming Riemann's hypothesis, that at least two thirds of the nontrivial zeroes are simple.

In 1974, Levinson proved that at least one third of the nontrivial zeroes of Riemann's zeta function are on the critical line. More precisely, if T is sufficiently large, $L = \log(T/2\pi)$, and $U = T/L^{10}$, then

$$N_0(T + U) - N_0(T) > \frac{1}{3}(N(T + U) - N(T)).$$

In 1989, Conrey has improved this result, replacing $\frac{1}{3}$ by $\frac{2}{5}$.

Extensive computations of the zeroes of $\zeta(s)$ have been made. These began with Gram in 1903, who computed the first 15 zeroes (that is, ρ_n for $1 \leq n \leq 15$). With the advent of computers, Rosser, Yohe & Schoenfeld had determined in 1969 the first 3,500,000 zeroes. Since then, this has been largely extended.

RECORD

Van de Lune, te Riele & Winter have determined (in 1986) that the first 1,500,000,001 nontrivial zeroes of $\zeta(s)$ are all simple, lie on the critical line, and have imaginary part with $0 < t < 545{,}439{,}823.215$. This work has involved over 1000 hours on a supercomputer.

Just not to be shamefully absent from this book, here is a table with the smallest zeroes $\rho_n = \frac{1}{2} + it_n$, $t_n > 0$:

n	t_n	n	t_n	n	t_n
1	14.134725	11	52.970321	21	79.337375
2	21.022040	12	56.446248	22	82.910381
3	25.010858	13	59.347044	23	84.735493
4	30.424876	14	60.831779	24	87.425275
5	32.935062	15	65.112544	25	88.809111
6	37.586178	16	67.079811	26	92.491899
7	40.918719	17	69.546402	27	94.651344
8	43.327073	18	72.067158	28	95.870634
9	48.005151	19	75.704691	29	98.831194
10	49.773832	20	77.144840	30	101.317851

In Edwards' book (1974), there is a detailed explanation of the method used by Gram, Backlund, Hutchinson, and Haselgrove to compute the smallest 300 zeroes of $\zeta(s)$. In 1986, Wagon wrote a short account with the essential information.

G. Zero-Free Regions for $\zeta(s)$ and the Error Term in the Prime Number Theorem

The knowledge of larger zero-free regions for $\zeta(s)$ leads to better estimates of the various functions connected with the distribution of primes.

I have already indicated that de la Vallée Poussin determined a zero-free region, which he used in an essential way in his proof of the prime number theorem. There have been many extensions of his result, and the largest known zero-free region has been determined by Richert [and published in Walfisz's book (1963)]:

It is to be remarked that up to now, no one has succeeded in knowing that $\zeta(s)$ has a zero-free region of the form $\{\, \sigma + it \mid \sigma \geq \sigma_0 \,\}$ with $\frac{1}{2} < \sigma_0 < 1$.

Using whatever is known about zero-free regions for $\zeta(s)$ it is possible to deduce an estimate for the error in the prime number theorem. Thus, Tschudakoff obtained

$$\pi(x) = \mathrm{Li}(x) + O\left(x e^{-C(\log x)^{\alpha}}\right);$$

with $\alpha < 4/7$, and $C > 0$.

In 1901, von Koch showed that Riemann's hypothesis is equivalent to the following form of the error

$$\pi(x) = \mathrm{Li}(x) + O(x^{1/2} \log x).$$

The knowledge that many zeroes of $\zeta(s)$ are on the critical line leads also to better estimates. Thus, Rosser & Schoenfeld proved in 1975 that

$$0.998684x < \theta(x) < 1.001102x$$

(the lower inequality for $x \geq 1319007$, the upper inequality for all x).

II. The nth Prime and Gaps

The results in the preceding section concern the asymptotic behaviour of $\pi(x)$—and its comparison with other known functions. But nothing was said about the behaviour of $\pi(x)$ in short intervals, nor about the size of the nth prime, the difference between consecutive primes, etc. These are questions putting in evidence the fine points in the distribution of primes, and much more irregularity is expected.

A. SOME PROPERTIES OF $\pi(x)$

In this respect, historically the first statement is Bertrand's experimental observation (in 1845):

Between $n \geq 2$ and $2n$ there is always a prime number.

Equivalently, this may be stated as

$$\pi(2n) - \pi(n) \geq 1 \qquad \text{(for } n \geq 2\text{),}$$

or also as

$$p_{n+1} < 2p_n \qquad \text{(for } n \geq 1\text{).}$$

This statement has been known as "Bertrand's postulate" and it was proved by Tschebycheff in 1852 as a by-product of his estimates already indicated for $\pi(x)$. As a matter of fact, the following inequalities are more refined:

$$1 < \frac{1}{3} \frac{n}{\log n} < \pi(2n) - \pi(n) < \frac{7}{5} \frac{n}{\log n} \qquad \text{(for } n \geq 5\text{),}$$

and clearly $\pi(4) - \pi(2) = 1$, $\pi(6) - \pi(3) = 1$, $\pi(8) - \pi(4) = 2$.

The simplest elementary proof of Bertrand's postulate, which I came across, is by Moser (1949).

More generally, Erdös proved in 1949 that for every $\lambda > 1$ there exists $C = C(\lambda) > 0$, and $x_0 = x_0(\lambda) > 1$ such that

$$\pi(\lambda x) - \pi(x) > C \frac{x}{\log x},$$

which is just a corollary of the prime number theorem.

The following result of Ishikawa (1934) is also a consequence of Tschebycheff's theorem:

If $x \geq y \geq 2$, $x \geq 6$, then $\pi(xy) > \pi(x) + \pi(y)$.

On the other hand, using deeper methods, Vaughan proved that

$$\pi(x + y) \leq \pi(x) + \frac{2y}{\log y};$$

by the result of Rosser & Schoenfeld (1975) already mentioned, it follows that $\pi(x + y) \leq \pi(x) + 2\pi(y)$.

The relation $\pi(2x) < 2\pi(x)$ for $x \geq 11$, was also proved by Rosser & Schoenfeld, as a consequence of their refined estimates of $\theta(x)$.

The following statements are still waiting to be proved, or disproved:

In 1882, Opperman stated that $\pi(n^2 + n) > \pi(n^2) > \pi(n^2 - n)$ for $n > 1$.

In 1904, Brocard asserted that $\pi(p_{n+1}^2) - \pi(p_n^2) \geq 4$ for $n \geq 2$; that is, between the squares of two successive primes greater than 2 there are at least four primes.

B. THE nTH PRIME

Now I shall consider specifically the nth prime.

The prime number theorem yields easily:

$$p_n \sim n \log n, \quad \text{that is,} \quad \lim_{n \to \infty} \frac{p_n}{n \log n} = 1.$$

In other words, for large indices n, the nth prime is about the size of $n(\log n$. More precisely

$$p_n = n \log n + n(\log \log n - 1) + o\left(\frac{n \log \log n}{\log n}\right).$$

So, for large n, $p_n > n \log n$. But Rosser proved in 1938 that for every $n > 1$:

$$n \log n + n(\log \log n - 10) < p_n < n \log n + n(\log \log n + 8)$$

and also that for every $n \geq 1$: $p_n > n \log n$.

The following results by Ishikawa (1934) are also consequences of Tschebycheff's theorems (see Trost's book):

If $n \geq 2$, then $p_n + p_{n+1} > p_{n+2}$;

if $m, n \geq 1$, then $p_m p_n > p_{m+n}$.

In a very interesting paper, Pomerance considered in 1979 the prime number graph, consisting of all the points (n, p_n) of the plane (with $n \geq 1$). He proved Selfridge's conjecture: there exist infinitely many n with $p_n^2 > p_{n-i} p_{n+i}$ for all positive $i < n$. Also, there are infinitely many n such that $2p_n < p_{n-i} + p_{n+i}$ for all positive $i < n$.

C. GAPS BETWEEN PRIMES

It is important to study the difference $d_n = p_{n+1} - p_n$ between consecutive primes, that is, the size of the gaps between consecutive primes. This gives an indication of the distribution of primes in short intervals.

It is quite easy to show that $\limsup d_n = \infty$, that is, for every $N > 1$ there exists a string of at least N consecutive composite integers; for example:

$$(N+1)! + 2, \ (N+1)! + 3, \ (N+1)! + 4, \ldots, \ (N+1)! + (N+1).$$

Actually, gaps of size N have been found experimentally between numbers much smaller than $(N+1)! + 1$.

For every $d \geq 1$ let $p(d)$ be the smallest prime following a string of d or more consecutive composite numbers.

RECORD

The largest d for which $p(d)$ has been calculated is $d = 804$, and $p(804) = 90874329412297$. This was determined by Young & Potler (1989).

To find $\liminf d_n$ is quite another matter—a difficult problem which I shall discuss in more detail (see (III)). It is not even known if $\liminf d_n < \infty$; this would mean that there exist some integer $k > 0$ such that there exist infinitely many pairs of successive primes with difference $2k$ (see Polignac's conjecture, in the subsection on possible gaps between primes).

The next natural question is to estimate the rate of growth of the difference $d_n = p_{n+1} - p_n$, and compare it with various functions of p_n. For example, to compare d_n with powers p_n^θ, (where $\theta > 0$).

For example, the result of Tschebycheff on Bertrand's postulate tells that $d_n < p_n$ for every $n \geq 1$.

By the prime number theorem,

$$\lim_{n \to \infty} \frac{d_n}{p_n} = 0.$$

Clearly, since $d_n < p_n$ for every $n \geq 1$, then $d_n = O(p_n)$.

Using Riemann's hypothesis, Cramér showed in 1937 that $d_n = O(p_n^{\frac{1}{2}} \log p_n)$.

Mathematicians are hoping to prove, without assuming Riemann's hypothesis, that $d_n = O(p_n^{\frac{1}{2}+\varepsilon})$ for every $\varepsilon > 0$. The race to reach this bound has been tight, from the first paper of Hoheisel (1930), with $d_n = O(p_n^\theta)$ and θ just below 1, through papers by Ingham (1937), Montgomery (1969), Huxley (1972), Iwaniec & Jutila (1979), Heath–Brown & Iwaniec (1979), Iwaniec & Pintz (1984), until the most recent records.

RECORD

In 1986, Mozzochi reached $\theta = \frac{1051}{1920}$ while Lou & Yao could go slightly lower, $\theta = \frac{85}{164}$ (in 1985).

Here is another difficult open problem: to show that

$$\lim_{n \to \infty} \left(\sqrt{p_{n+1}} - \sqrt{p_n} \right) = 0.$$

If true, this would establish, for n sufficiently large, the conjecture of D. Andrica (1978) namely, $\sqrt{p_{n+1}} - \sqrt{p_n} < 1$, which has been checked for $p_n < 10^6$.

In turn, if the latter conjecture holds, it would follow that between the squares of any two consecutive integers, there is always a prime number, a seemingly true fact which is as yet unproved. Note that this is weaker than one of the assertions conjectured by Opperman.

D. The Possible Gaps between Primes

Now that the growth of d_n has been discussed, it is also natural to ask which values are possible for d_n; of course, for $n > 1$, $d_n = p_{n+1} - p_n$ is even.

The following result is an easy application of the prime number theorem, and was proposed by Powell as a problem in the *American Mathematical Monthly* (1983; solution by Davies in 1984):

For every natural number M, there exists an even number $2k$, such that there are more than M pairs of successive primes with difference $2k$.

Proof. Let n be sufficiently large, consider the sequence of primes

$$3 = p_2 < p_3 < \cdots < p_n$$

and the $n - 2$ differences $p_{i+1} - p_i$ $(i = 2, \ldots, n-1)$. If the number of distinct differences is less than

$$\left[\frac{n-2}{M} \right],$$

then one of these differences, say $2k$, would appear more than M times. In the alternative case,

$$p_n - p_2 \geq 2 + 4 + \cdots + 2 \left[\frac{n-2}{M} \right].$$

But the right-hand side is asymptotically equal to n^2/M^2, while the left-hand side is asymptotically equal to $n \log n$, by the prime number theorem. This is impossible. $\qquad\square$

Of a totally different order of difficulty is Polignac's conjecture (1849): for every even natural number $2k$ there are infinitely many pairs of consecutive primes p_n, p_{n+1} such that $d_n = p_{n+1} - p_n = 2k$.

In particular, this conjecture includes as a special case the following open problem: are there infinitely many pairs of primes p, $p + 2$?

A positive answer means that $\liminf d_n = 2$, where $d_n = p_{n+1} - p_n$.

This question will be considered in the next section. I remark in this connection, that it is also not known that every even natural number is a difference of two primes (even without requiring them to be consecutive).

III. Twin Primes

If p and $p + 2$ are primes, they are called twin primes.

The smallest pairs of twin primes are $(3,5)$, $(5,7)$, $(11,13)$ and $(17,19)$. Twin primes have been characterized by Clement in 1949 as follows.

Let $n \geq 2$. The integers n, $n + 2$ form a pair of twin primes if and only if

$$4[(n - 1)! + 1] + n \equiv 0 \pmod{n(n + 2)}.$$

Proof. If the congruence is satisfied, then $n \neq 2$, 4, and

$$(n - 1)! + 1 \equiv 0 \pmod{n},$$

so by Wilson's theorem n is a prime. Also

$$4(n - 1)! + 2 \equiv 0 \pmod{n + 2};$$

hence multiplying by $n(n + 1)$,

$$4[(n + 1)! + 1] + 2n^2 + 2n - 4 \equiv 0 \pmod{n + 2}$$

and then

$$4[(n + 1)! + 1] + (n + 2)(2n - 2) \equiv 0 \pmod{n + 2};$$

from Wilson's theorem $n + 2$ is also a prime.

Conversely, if n, $n + 2$ are primes, then $n \neq 2$ and

$$
\begin{aligned}
(n - 1)! + 1 &\equiv 0 \pmod{n}, \\
(n + 1)! + 1 &\equiv 0 \pmod{n + 2}.
\end{aligned}
$$

But $n(n + 1) = (n + 2)(n - 1) + 2$, so $2(n - 1)! + 1 = k(n + 2)$, where k is an integer. From $(n - 1)! \equiv -1 \pmod{n}$, then $2k + 1 \equiv 0 \pmod{n}$ and substituting $4(n - 1)! + 2 \equiv -(n + 2) \pmod{n(n + 2)}$, therefore $4[(n - 1)! + 1] + n \equiv 0 \pmod{n(n + 2)}$. $\qquad\square$

However, this characterization has no practical value in the determination of twin primes.

The main problems is to ascertain whether there exist infinitely many twin primes.

For every $x > 1$, let $\pi_2(x)$ denote the number of primes p such that $p + 2$ is also prime and $p + 2 \leq x$.

Brun announced in 1919 that there exists an effectively computable integer x_0 such that, if $x \geq x_0$, then

$$\pi_2(x) < \frac{100x}{(\log x)^2}.$$

The proof appeared in 1920.

In another paper of 1919, Brun proved the famous result that

$$\sum \left(\frac{1}{p} + \frac{1}{p+2} \right) \text{ (sum for all primes } p \text{ such that } p + 2 \text{ is also a prime)}$$

converges, which expresses the scarcity of twin primes, even if there are infinitely many of them. The sum

$$B = \left(\frac{1}{3} + \frac{1}{5} \right) + \left(\frac{1}{5} + \frac{1}{7} \right) + \left(\frac{1}{11} + \frac{1}{13} \right) + \cdots + \left(\frac{1}{p} + \frac{1}{p+2} \right) + \cdots$$

is now called *Brun's constant*. Based on heuristic considerations about the distribution of twin primes, B has been calculated, for example, by Shanks & Wrench (1974) and by Brent (1976):

$$B = 1.90216054 \ldots.$$

Brun also proved that for every $m \geq 1$ there exist m successive primes which are not twin primes.

The estimate of $\pi_2(x)$ has been refined by a determination of the constant and of the size of the error. This was done, among others, by Bombieri & Davenport, in 1966. It is an application of sieve methods and its proof may be found, for example, in the book of Halberstam & Richert.

Here is the result:

$$\pi_2(x) \leq 2C \prod_{p>2} \left(1 - \frac{1}{(p-1)^2} \right) \frac{x}{(\log x)^2};$$

Hardy & Littlewood (1923) conjectured that the constant C is equal to 1. The best values obtained thus far for C have been: $C = 3.5$ by Bombieri, Friedlander & Iwaniec (1986) and $C = 3.13$ by S. Lou (still unpublished).

The infinite product

$$C_2 = \prod_{p>2} \left(1 - \frac{1}{(p-1)^2}\right)$$

is called the *twin-prime constant*; its value is 0.66016.... It had been calculated by Wrench in 1961.

To give a feeling for the growth of $\pi_2(x)$, I reproduce below part of the calculations of Brent (1975, 1976):

x	$\pi_2(x)$
10^3	35
10^4	205
10^5	1224
10^6	8169
10^7	58980
10^8	440312
10^9	3424506
10^{10}	27412679
10^{11}	224376048

RECORD

The largest exact value for the number of twin primes below a given limit has been published by Brent in 1976:

$$\pi_2(10^{11}) = 224{,}376{,}048.$$

RECORD

The largest known pairs of twin primes are $1706595 \times 2^{11235} \pm 1$ and $571305 \times 2^{7701} \pm 1$, found in 1990 by B. Parady, J. Smith, and S. Zarantonello. Previously, large pairs of twin primes, also with more than 1000 digits, were found by H. Dubner, O. A. L. Atkin, N. W. Rickert, and W. Keller.

Sieve theory has been used in attempts to prove that there exists infinitely many twin primes. Many authors have worked with this method.

To begin, in his famous paper of 1920, Brun showed that 2 may be written, in infinitely many ways, in the form $2 = m - n$, where m, n are products of at most 9 primes (not necessarily distinct).

The best result to date, is due to Chen (announced in 1966, published in 1973, 1978); he proved that 2 may be written in infinitely many ways in the form $2 = m - p$, with p prime, and m a product of at most two primes (not necessarily distinct).

The sieve methods used for the study of twin primes are also appropriate for the investigation of Goldbach's conjecture (see Section VI).

ADDENDUM ON POLIGNAC'S CONJECTURE

The general Polignac's conjecture can be, in part, treated like the twin-primes conjecture.

For every $k \geq 1$ and $x > 1$, let $\pi_{2k}(x)$ denote the number of integers $n > 1$ such that $p_{n+1} \leq x$ and $p_{n+1} - p_n = 2k$.

With Brun's method, it may be shown that there exists a constant $C_k > 0$ such that

$$\pi_{2k}(x) < C_k \frac{x}{(\log x)^2}.$$

IV. Primes in Arithmetic Progression

A. THERE ARE INFINITELY MANY!

A classical and most important theorem was proved by Dirichlet in 1837. It states:

If $d \geq 2$ and $a \neq 0$ are integers that are relatively prime, then the arithmetic progression

$$a, \ a+d, \ a+2d, \ a+3d, \ldots$$

contains infinitely many primes.

Many special cases of this theorem were already known including of course Euclid's theorem (when $a = 1$, $d = 2$).

Indeed, if $d = 4$ or $d = 6$ and $a = -1$, the proof is very similar to Euclid's proof.

Using simple properties of quadratic residues, it is also easy to show that each of these arithmetic progressions contain infinitely many primes:

$d = 4$, $a = 1$;
$d = 6$, $a = 1$;

$d = 3$, $a = 1$;

$d = 8$, $a = 3$, or $a = 5$, or $a = 7$ (this includes the progressions with $d = 4$);

$d = 12$, $a = 5$, or $a = 7$, or $a = 11$ (this includes also the progressions with $d = 6$).

For $d = 8$, 16, or more generally $d = 2^r$, and $a = 1$, the ingredients of a simple proof are: to consider $f(N)$ where

$$f(X) = X^{2^{r-1}} + 1, \quad N = 2p_1p_2 \cdots p_n,$$

each p_i prime, $p_i \equiv 1 \pmod{2^r}$, and then to use Fermat's little theorem. These are hints for a reader wanting to find this proof by himself.

The proof when d is arbitrary and $a = 1$ or $a = -1$ is also elementary, though not so simple, and requires the cyclotomic polynomials and some of their elementary properties.

A detailed discussion of Dirichlet's theorem, with several variants of proofs, is in Hasse's book (1950) *Vorlesungen über Zahlentheorie* (now available in English translation, from Springer-Verlag).

In 1949, Selberg gave an elementary proof of Dirichlet's theorem, similar to his proof of the prime number theorem.

Concerning Dirichlet's theorem, de la Vallée Poussin established the following additional density result. For a, d as before, and $x \geq 1$ let

$$\pi_{d,a}(x) = \#\{\, p \text{ prime} \mid p \leq x, \, p \equiv a \pmod{d} \,\}.$$

Then

$$\pi_{d,a}(x) \sim \frac{1}{\phi(d)} \cdot \frac{x}{\log x}.$$

Note that the right-hand side is the same, for any a, such that $\gcd(a, d) = 1$.

It follows that

$$\lim_{x \to \infty} \frac{\pi_{d,a}(x)}{\pi(x)} = \frac{1}{\phi(d)},$$

and this may be stated by saying that the set of primes in the arithmetic progression $\{\, a + kd \mid k \geq 1 \,\}$ has natural density $1/\phi(d)$ (with respect to the set of all primes).

Despite the fact that the asymptotic behavior of $\pi_{d,a}(x)$ is the same, for every a, $1 \leq a < d$, with $\gcd(a, d) = 1$, Tschebycheff had noted, already in 1853, that $\pi_{3,1}(x) < \pi_{3,2}(x)$ and $\pi_{4,1}(x) < \pi_{4,3}(x)$ for small values of x; in other words, there are more primes of the form $3k + 2$ than of the form $3k + 1$ (resp. more primes $4k + 3$ than primes $4k + 1$) up to x (for x not too large). Are these inequalities true for every x? The situation is somewhat similar to that of the inequality $\pi(x) < \mathrm{Li}(x)$. Once again, as in Littlewood's theorem, it may be shown that these inequalities are reversed infinitely often. Thus, Leech computed in 1957 that $x_1 = 26861$ is the smallest prime for which $\pi_{4,1}(x) > \pi_{4,3}(x)$; see also Bays & Hudson (1978), who found that $x_1 = 608,981,813,029$ is the smallest prime for which $\pi_{3,1}(x) > \pi_{3,2}(x)$.

For the computation of the exact number of primes $\pi_{d,a}(x)$, below x in the arithmetic progression $\{\, a + kd \mid k \geq 1 \,\}$, Hudson derived in 1977, a formula similar to Meissel's formula for the exact value of $\pi(x)$. In the same year, Hudson & Brauer studied in more detail the particular arithmetic progressions $4k \pm 1$, $6k \pm 1$.

B. THE SMALLEST PRIME IN AN ARITHMETIC PROGRESSION

With $d \geq 2$ and $a \geq 1$, relatively prime, let $p(d, a)$ be the smallest prime in the arithmetic progression $\{\, a + kd \mid k \geq 0 \,\}$. Can one find an upper bound, depending only on a, d, for $p(d, a)$?

Let $p(d) = \max\{\, p(d, a) \mid 1 \leq a < d, \gcd(a, d) = 1 \,\}$. Again, can one find an upper bound for $p(d)$ depending only on d? And how about lower bounds?

Linnik's theorem of 1944, which is one of the deepest results in analytic number theory asserts:

There exists $d_0 \geq 2$ and $L > 1$ such that $p(d) < d^L$ for every $d \geq d_0$.

Note that the absolute constant L, called *Linnik's constant*, is effectively computable.

It is clearly important to compute the value of L. Pan (Cheng-Dong) was the first to evaluate Linnik's constant, giving $L \leq 5448$ in 1957 (see 1965). Subsequently, a number of papers have appeared, where the estimate of the constant was improved.

RECORD

Heath–Brown (1989) has shown that $L \leq 6.5$; this paper is still unpublished. Previous records were due to Chen, Jutila, Graham.

Schinzel & Sierpiński (1958) and Kanold (1963) have conjectured that $L = 2$, that is, $p(d) < d^2$ for every sufficiently large $d \geq 2$.

Explicitly, this means that if $1 \leq a < d$, $\gcd(a, d) = 1$, there is a prime number among the numbers a, $a + d$, $a + 2d$, ..., $a + (d - 1)d$.

Heath-Brown has advanced in 1978 the conjecture that $p(d) \leq Cd(\log d)^2$ and Wagstaff sustained in 1979 that $p(d) \sim \phi(d)(\log d)^2$ on heuristic grounds.

Concerning the lower bounds for $p(d)$, I first mention the following result of Schatunowsky (1893), obtained independently by Wolfskehl[*] *Paul Wolfskehl is usually remembered as the rich mathematician who endowed a substantial prize for the discovery of a proof of Fermat's last theorem. I tell the story in my book *13 Lectures on Fermat's Last Theorem*. Here, I wish to salute his memory, on behalf of all the young assistants, who in the last 80 years, have been thrilled finding mistakes in an endless and continuous flow of "proofs" of Fermat's theorem. in 1901.

$d = 30$ is the large integer with the following property: if $1 \leq a < d$ and $\gcd(a, d) = 1$, then $a = 1$ or a is prime.

The proof is elementary, and may be found in Landau's book *Primzahlen* (1909), page 229. It follows at once that if $d > 30$ then $p(d) > d + 1$.

From the prime number theorem, it follows already that for every $\varepsilon > 0$ and for all sufficiently large d:

$$p(d) > (1 - \varepsilon)\phi(d)\log d.$$

This implies that

$$\liminf \frac{p(d)}{\phi(d)\log d} \geq 1.$$

Moreover, $\frac{p(d)}{\phi(d)\log d}$ is unbounded.

C. Strings of Primes in Arithmetic Progression

Now I consider the question of existence of sequences of k primes $p_1 < p_2 < p_3 < \cdots < p_k$ with difference $p_2 - p_1 = p_3 - p_2 = \cdots = p_k - p_{k-1}$, so these primes are in arithmetic progression.

In 1939, van der Corput proved that there exist infinitely many sequences of three primes in arithmetic progression (see Section VI).

Even though it is easy to give examples of four (and sometimes more) primes in arithmetic progression, the following question remains open:

Is it true that for every $k \geq 4$, there exist infinitely many arithmetic progressions consisting of k primes?

For the first case, $k = 4$, to date the best result was obtained by Heath-Brown, who showed in 1981 that there exist infinitely many arithmetic progressions consisting of four numbers, of which three are primes and the other is the product of two, not necessarily distinct, prime factors.

It is conjectured that for every $k > 3$ there exists at least one arithmetic progression consisting of k prime numbers.

There have been extensive computer searches for long strings of primes in arithmetic progression.

Record

The longest known string of primes in arithmetic progression contains 20 primes, of which the smallest is $p = 214861583621$ and the difference is $d = 1884649770$. This string has been discovered by Young & Fry in 1987.

The previous records were by Pritchard (19 terms in 1985, 18 terms in 1982) and Weintraub (17 terms in 1977). Much computation is required in the search of long strings of primes in arithmetic progression.

In this connection the following statement, due to M. Cantor (1861), and quoted in Dickson's *History of the Theory of Numbers*, Vol. I, p. 425, is easy to prove:

Let $d \geq 2$, let a, $a + d$, ..., $a + (n - 1)d$ be n prime numbers in arithmetic progression. Let q be the largest prime such that $q \leq n$. Then either $\prod_{p \leq q} p$ divides d, or $a = q$ and $\prod_{p < q} p$ divides d.

Proof. First an easy remark. If p is a prime, not dividing d, and if a, $a+d$, ..., $a+(p-1)a$ are primes, then these numbers are pairwise incongruent modulo p and p divides exactly one of these numbers. Now assume that $\prod_{p \leq q} p$ does not divide d, so that there exists a prime $p \leq n$ such that p does not divide d. Choose the smallest such prime p. By the remark, there exists j, $0 \leq j \leq p - 1$, such that p divides $a + jd$, so $p = a + jd$, since $a + jd$ is a prime number. But a is a prime; if $a \neq a + jd$, then a divides d (by the choice of p), so a divides p, that is $a = p = a + jd$. This proves that $p = a$. If $p < q$, then $p \leq n - 1$, so p divides $a + pd$, hence $p = a + pd = p(1 + d)$, which is an absurdity.

I have therefore established that if $\prod_{p \leq q} p$ does not divide d then $q = a$ and $\prod_{p < q} p$ divides d. $\qquad\qquad\qquad\qquad\qquad\qquad\qquad\square$

A particular case of this proposition had been established by Lagrange.

At this point, I recall that in Chapter III, Section II, I have discussed the even more difficult question of finding strings of p primes in arithmetic progression, of which the smallest number is equal to p itself.

A related conjecture is the following:

There exist arbitrarily long arithmetic progressions of *consecutive* primes.

RECORD

The longest known string of consecutive primes in arithmetic progression contains six terms; its difference is 30 and the initial term is 121174811. It was discovered by Lander & Parkin (1967). A second example was discovered by Weintraub (1977); its difference is 30 and it begins with 999,900,067,719,989.

V. Goldbach's Famous Conjecture

In a letter of 1742 to Euler, Goldbach expressed the belief that:

(G) *Every integer $n > 5$ is the sum of three primes.*

Euler replied that this is easily seen to be equivalent to the following statement:

(G′) *Every even integer $2n \geq 4$ is the sum of two primes.*

Indeed, if (G′) is assumed to be true and if $2n \geq 6$, then $2n - 2 = p + p'$ so $2n = 2 + p + p'$, where p, p' are primes. Also $2n + 1 = 3 + p + p'$, which proves (G).

Conversely, if (G) is assumed to be true, and if $2n \geq 4$, then $2n + 2 = p + p' + p''$, with p, p', p'' primes; then necessarily $p'' = 2$ (say) and $2n = p + p'$.

Note that it is trivial that (G′) is true for infinitely many even integers: $2p = p + p$ (for every prime).

Very little progress was made in the study of this conjecture before the development of refined analytical methods and sieve theory. And despite all the attempts, the problem is still unsolved.

There have been three main lines of attack, reflected, perhaps inadequately, by the keywords "asymptotic," "almost primes," "basis."

An asymptotic statement is one which is true for all sufficiently large integers.

The first important result is due to Hardy & Littlewood in 1923—it is an asymptotic theorem. Using the circle method and a modified form of the Riemann's hypothesis, they proved that there exists n_0 such that every odd number $n \geq n_0$ is the sum of three primes.

Later, in 1937, Vinogradov gave a proof of Hardy & Littlewood's theorem, without any appeal to the Riemann hypothesis. There have been calculations of n_0, which may be taken to be $n_0 = 3^{3^{15}}$.

A natural number $n = \prod_{i=1}^{r} p_i^{e_i}$ (where each p_i is a prime) is called a *k-almost prime* when $\sum_{i=1}^{r} e_i = k$; the set of k-almost primes is denoted by P_k.

The approach via almost-primes consists in showing that there exist $h, k \geq 1$ such that every sufficiently large even integer is in the set $P_h + P_k$ of sums of integers of P_h and of P_k. What is intended is, of course, to show that h, k can be taken to be 1.

In this direction, the first result is due to Brun (1919, *C.R. Acad. Sci. Paris*): every sufficiently large even number belongs to $P_9 + P_9$.

Much progress has been achieved, using more involved types of sieve. In 1950, Selberg showed that every sufficiently large even integer is in $P_2 + P_3$.

While these results involved summands which were both composite, Rényi proved in 1947 that there exists an integer $k \geq 1$ such that every sufficiently large even integer is in $P_1 + P_k$. Subsequent work provided explicit values of k.

The best result to date—and the closest one has come to establishing Goldbach's conjecture—is by Chen (announcement of results in 1966; proofs in detail in 1973, 1978). In his famous paper, Chen proved:

Every sufficiently large even integer may be written as $2n = p + m$, where p is a prime and $m \in P_2$.

As I mentioned before, Chen proved at the same time, the "conjugate" result that there are infinitely many primes p such that $p + 2 \in P_2$; this is very close to showing that there are infinitely many twin primes.

The same method is good to show that for every even integer $2k \geq 2$, there are infinitely many primes p such that $p + 2k \in P_2$; so $2k$ is the difference $m - p$ ($m \in P_2$, p prime) in infinitely many ways.

A proof of Chen's theorem is given in the books of Halberstam & Richert and Wang (1984). See also the simpler proof given by Ross (1975).

The "basis" approach began with the famous theorem of Schnirelmann (1930), proved, for example, in Landau's book (1937) and in Gelfond & Linnik's book (translated in 1965):

There exists a positive integer S, such that every sufficiently large integer is the sum of at most S primes.

It follows that there exist a positive integer $S_0 \geq S$ such that every integer (greater than 1) is a sum of at most S_0 primes.

S_0 is called the *Schnirelmann constant*.

In his small and neat book (1947), Khinchin wrote an interesting and accessible chapter on Schnirelmann's ideas of bases and density of sequences of numbers.

Schnirelmann's constant S_0 has been effectively estimated.

RECORD

The best estimate for Schnirelmann's constant up to now is $S_0 = 19$, and it is due to Riesel & Vaughan (1983).

In 1949, Richert proved the following analogue of Schnirelmann's theorem: every integer $n > 6$ is the sum of distinct primes.

Here I note that Schinzel showed in 1959 that Goldbach's conjecture implies (and so, it is equivalent to) the statement:

Every integer $n > 17$ is the sum of exactly three distinct primes.

Thus, Richert's result will be a corollary of Goldbach's conjecture (if and when it will be shown true).

Now I shall deal with the number $r_2(2n)$ of representations of $2n \geq 4$ as sums of two primes. *A priori*, $r_2(2n)$ might be zero (until Goldbach's conjecture is established).

Hardy & Littlewood gave, in 1923, the asymptotic formula below, which at first relied on a modified Riemann's hypothesis; later work of Vinogradov removed this dependence:

$$r_2(2n) \le C \frac{2n}{(\log 2n)^2} \log \log 2n.$$

On the other hand, Powell proposed in 1985 to give an elementary proof of the following fact (problem in *Mathematics Magazine*): for every $k > 0$ there exist infinitely many even integers $2n$, such that $r_2(2n) > k$. A solution by Finn & Frohliger was published in 1986.

For every $x \ge 4$ let

$$G'(x) = \#\{ 2n \mid 2n \le x, \ 2n \text{ is not a sum of two primes} \}.$$

Van der Corput (1937), Estermann (1938), and Tschudakoff (1938) proved independently that $\lim G'(x)/x = 0$, and in fact, $G'(x) = O(x/(\log x)^\alpha)$, for every $\alpha > 0$.

The best result in this direction is the object of a deep paper by Montgomery & Vaughan (1975), and it asserts that there exists an effectively computable constant α, $0 < \alpha < 1$, such that for every sufficiently large x, $G'(x) < x^{1-\alpha}$. In 1980, Chen showed that $\alpha = \frac{1}{25}$ is a possible choice; this followed earlier work by Chen & Pan (1980).

Concerning numerical calculations about Goldbach's conjecture, I give now the records.

RECORD

In 1965, Stein & Stein verified Goldbach's conjecture up to 10^8 and calculated $r_2(2n)$ for every even number up to 200000; they conjectured that $r_2(2n)$ may take any integral value. Light, Forrest, Hammond & Roe, unaware of Stein & Stein's calculations, have again verified, in 1980, Goldbach's conjecture up to 10^8.

VI. The Distribution of Pseudoprimes and of Carmichael Numbers

Now I shall indicate results on the distribution of pseudoprimes and of Carmichael numbers.

A. DISTRIBUTION OF PSEUDOPRIMES

Let $P\pi(x)$ denote the number of pseudoprimes (to the base 2) less than or equal to x.

Let $(\mathrm{psp})_1 < (\mathrm{psp})_2 < \cdots < (\mathrm{psp})_n < \cdots$ be the increasing sequence of pseudoprimes.

In 1949 and 1950, Erdös gave the following estimates

$$C \log x < P\pi(x) < \frac{x}{e^{\frac{1}{3}(\log x)^{1/4}}}$$

(for x sufficiently large and $C > 0$). They were later to be much improved, as I shall soon indicate.

From these estimates, it is easy to deduce that

$$\sum_{n=1}^{\infty} 1/(\mathrm{psp})_n$$

is convergent (first proved by Szymiczek in 1967), while

$$\sum_{n=1}^{\infty} 1/\log(\mathrm{psp})_n$$

is divergent (first shown by Mąkowski in 1974).

It is convenient to introduce the following notation for the counting functions of pseudoprimes, Euler and strong pseudoprimes in arbitrary bases $a \geq 2$:

$$P\pi_a(x) = \#\{\, n \mid 1 \leq n \leq x,\ n \text{ is } \mathrm{psp}(a)\,\}, \qquad P\pi(x) = P\pi_2(x),$$
$$EP\pi_a(x) = \#\{\, n \mid 1 \leq n \leq x,\ n \text{ is } \mathrm{epsp}(a)\,\}, \qquad EP\pi(x) = EP\pi_2(x),$$
$$SP\pi_a(x) = \#\{\, n \mid 1 \leq n \leq x,\ n \text{ is } \mathrm{spsp}(a)\,\}, \qquad SP\pi(x) = SP\pi_2(x).$$

Clearly, $SP\pi_a(x) \leq EP\pi_a(x) \leq P\pi_a(x)$.

And now, the estimates of upper and lower bounds for these functions.

For the upper bound of $P\pi(x)$, improving on a previous result of Erdös (1956), Pomerance showed in 1981, that for all large x:

$$P\pi(x) \leq \frac{x}{\ell(x)^{1/2}},$$

with

$$\ell(x) = e^{\log x \log \log \log x / \log \log x}.$$

The same bound is also good for $P\pi_a(x)$, with arbitrary basis a.

Concerning lower bounds, the best result to date is by Pomerance, in 1982 (see Remark 3 of his paper):

$$e^{(\log x)^{\alpha}} \leq SP\pi_a(x),$$

where $\alpha = 5/14$.

The tables of pseudoprimes suggest that for every $x > 170$ there exists a pseudoprime between x and $2x$. However, this has not yet been proved. In this direction, I note the following results of Rotkiewicz (1965):

If n is an integer, $n > 19$, there exists a pseudoprime between n and n^2.

Also, for every $\varepsilon > 0$ there exists $x_0 = x_0(\varepsilon) > 0$ such that, if $x > x_0$, then there exists a pseudoprime between x and $x^{1+\varepsilon}$.

Concerning pseudoprimes in arithmetic progression, Rotkiewicz proved in 1963 and 1967:

If $a \geq 1$, $d \geq 1$ and $\gcd(a, d) = 1$, there exist infinitely many pseudoprimes in the arithmetic progression $\{\, a + kd \mid k \geq 1 \,\}$. Let $\mathrm{psp}(d, a)$ denote the smallest pseudoprime in this arithmetic progression. Rotkiewicz showed in 1972:

For every $\varepsilon > 0$ and for every sufficiently large d,

$$\log \mathrm{psp}(d, a) < d^{4L^2 + L + \varepsilon},$$

where L is Linnik's constant (seen in Section IV).

The above results have been extended by van der Poorten & Rotkiewicz in 1980: if a, $d \geq 1$, $\gcd(a, d) = 1$, then the arithmetic progression $\{\, a + kd \mid k \geq 1 \,\}$ contains infinitely many odd strong pseudoprimes for each base $b \geq 2$.

B. DISTRIBUTION OF CARMICHAEL NUMBERS

Now I turn my attention to the distribution of Carmichael numbers.

Let $CN(x)$ denote the number of Carmichael numbers n such that $n \leq x$.

I recall that it is believed that there exist infinitely many Carmichael numbers, but this assertion has still to be proved.

In 1956, Erdös showed that there exists a constant $\alpha > \frac{1}{2}$ such that, for every sufficiently large x,

$$CN(x) \leq \frac{x}{\ell(x)^\alpha},$$

where $L(x)$ was defined above.

Pomerance, Selfridge & Wagstaff improved this estimate in 1980: for every $\varepsilon > 0$ there exists $x_0(\varepsilon) > 0$ such that, if $x \geq x_0(\varepsilon)$, then

$$CN(x) \leq \frac{x}{\ell(x)^{1-\varepsilon}}.$$

It is much more difficult to provide lower bounds for $CN(x)$.

I refer now to tables of pseudoprimes, Carmichael numbers.

Poulet determined, already in 1938, all the (odd) pseudoprimes (in base 2) up to 10^8. Carmichael numbers were starred in Poulet's table.

In 1975, Swift compiled a table of Carmichael numbers up to 10^9.

The table of Pomerance, Selfridge & Wagstaff (1980) comprises pseudoprimes, Euler pseudoprimes, strong pseudoprimes (in base 2) and Carmichael numbers and extends up to 25×10^9. The largest table of Carmichael numbers, up to 10^{12}, is by Jaeschke (1990).

x	$P\pi(x)$	$EP\pi(x)$	$SP\pi(x)$	$CN(x)$
10^3	3	1	0	1
10^4	22	12	5	7
10^5	78	36	16	16
10^6	245	114	46	43
10^7	750	375	162	105
10^8	2057	1071	488	255
10^9	5597	2939	1282	646
10^{10}	14887	7706	3291	1547
25×10^9	21853	11347	4842	2163

From Jaeschke's table, $CN(10^{12}) = 8238$. If $CN(k, x)$ denotes the number of Carmichael numbers $n \leq x$ having k prime factors, Jaeschke obtained:

k	$CN(k, 10^{12})$
3	1000
4	2102
5	3156
6	1713
7	260
8	7
9	0

C. DISTRIBUTION OF LUCAS PSEUDOPRIMES

In Chapter 2, Section X the Lucas pseudoprimes were studied. Recall that if P, Q are nonzero integers, $D = P^2 - 4Q$, and the Lucas sequence is defined by

$$U_0 = 0, \quad U_1 = 1, \quad U_n = PU_{n-1} - QU_{n-2} \quad \text{(for } n \geq 2\text{)},$$

then the composite integer n, relatively prime to D, is a Lucas pseudoprime (with parameters (P, Q)) when n divides $U_{n-(D/n)}$.

Since the concept of Lucas pseudoprime is quite recent, much less
is known about the distribution of such numbers. My source is the
paper of Baillie & Wagstaff (1980) quoted in Chapter 2. Here are the
main results:

If x is sufficiently large, the number $L\pi(x)$ of Lucas pseudoprimes
(with parameters (P, Q)), $n \le x$, is bounded as follows:

$$L\pi(x) < \frac{x}{e^{Cs(x)}}$$

where $C > 0$ is a constant and $s(x) = (\log x \log \log x)^{1/2}$.

It follows (like in Szymiczek's result for pseudoprimes) for any
given parameters (P, Q), $\sum(1/U_n)$ is convergent (sum for all Lucas
pseudoprimes with these parameters).

On the other hand, in 1988, Erdős, Kiss & Sárközy showed that
there is a constant $C > 0$ such that for any nondegenerate Lucas
sequence, and x sufficiently large, $L\pi(x) > \exp\{(\log x)^C\}$.

Similarly, there is the following lower bound for the number $SL\pi(x)$
of strong Lucas pseudoprimes (with parameters (P, Q)) $n \le x$ (see
the definition in Chapter 2, Section X):

$SL\pi(x) > C' \log x$ (valid for all x sufficiently large) where $C' > 0$
is a constant.

5

Which Special Kinds of Primes Have Been Considered?

We have already encountered several special kinds of primes, for example, those which are Fermat numbers, or Mersenne numbers (see Chapter 2). Now I shall discuss other families of primes, among them the regular primes, the Sophie Germain primes, the Wieferich primes, the Wilson primes, the prime repunits, the primes in second-order linear recurring sequences.

Regular primes, Sophie Germain primes, and Wieferich primes have directly sprung from attempts to prove Fermat's last theorem. The interested reader may wish to consult my book *13 Lectures on Fermat's Last Theorem*, where these matters are considered in more detail. In particular, there is an ample bibliography including many classical papers, which will not be listed in the bibliography of this book.

I. Regular Primes

Regular primes appeared in the work of Kummer in relation with Fermat's last theorem. In a letter to Liouville, in 1847, Kummer stated that he had succeeded in proving Fermat's last theorem for all primes p satisfying two conditions. Indeed, he showed that if p satisfies these conditions, then there do not exist integers x, y, $z \neq 0$ such that $x^p + y^p = z^p$. He went on saying that "it remains only to find out whether these properties are shared by all prime numbers."

To describe these properties I need to explain some concepts that were first introduced by Kummer.

Let p be an odd prime, let

$$\zeta = \zeta_p = \cos \frac{2\pi}{p} + i \sin \frac{2\pi}{p}$$

be a primitive pth root of 1. Note that $\zeta^{p-1} + \zeta^{p-2} + \cdots + \zeta + 1 = 0$, because $X^p - 1 = (X - 1)(X^{p-1} + X^{p-2} + \cdots + X + 1)$ and $\zeta^p = 1$, $\zeta \neq 1$. Thus, ζ^{p-1} is expressible in terms of the lower powers of ζ. Let K be the set of all numbers $a_0 + a_1\zeta + \cdots + a_{p-2}\zeta^{p-2}$, with a_0, $\ldots$, a_{p-2} rational numbers. Let A be the subset of K consisting of those numbers for which $a_0, a_1, \ldots, a_{p-2}$ are integers. Then K is a field, called the field of $(p\text{-})$ cyclotomic numbers. A is a ring, called the ring of $(p\text{-})$ cyclotomic integers. The units of A are the numbers $\alpha \in A$ which divide 1, that is $\alpha\beta = 1$ for some $\beta \in A$. The element $\alpha \in A$ is called a cyclotomic prime if α cannot be written in the form $\alpha = \beta\gamma$ with $\beta, \gamma \in A$, unless β or γ is a unit.

I shall say that the arithmetic of p - cyclotomic integers is ordinary if every cyclotomic integer is equal, in unique way, up to units, to the product of cyclotomic primes.

Kummer discovered as early as 1847, that if $p \leq 19$ then the arithmetic of the p-cyclotomic integers is ordinary; however it is not so for $p = 23$.

To find a way to deal with nonuniqueness of factorization, Kummer introduced ideal numbers. Later, Dedekind considered certain sets of cyclotomic integers, which he called ideals. I'm refraining from defining the concept of ideal, assuming that it is well known to the reader. Dedekind ideals provided a concrete description of Kummer ideal numbers, so it is convenient to state Kummer's results in terms of Dedekind ideals. A prime ideal P is an ideal which is not equal to 0 nor to the ring A, and which cannot be equal to the product of two ideals, $P = IJ$, unless I or J is equal to P. Kummer showed that for every prime $p > 2$, every ideal of the ring of p-cyclotomic integers is equal, in a unique way, to the product of prime ideals.

In this context, it was natural to consider two nonzero ideals I, J to be equivalent if there exist nonzero cyclotomic integers α, $\beta \in A$ such that $A\alpha.I = A\beta.J$. The set of equivalence classes of ideals forms a commutative semigroup, in which the cancellation property holds; there is therefore a smallest group containing it. This is called the group of ideal classes. Kummer showed that this group is finite. The number of its elements is called the class number, and denoted by $h = h(p)$. It is a very important arithmetic invariant. The class number $h(p)$ is equal to 1 exactly when every ideal of A is principal, that is, of the form $A\alpha$, for some $\alpha \in A$. Thus $h(p) = 1$ exactly when the arithmetic of the p-cyclotomic integers is ordinary. So, the size of $h(p)$ is one measure of the deviation from the ordinary arithmetic.

Let it be said here that Kummer developed a very deep theory, obtained an explicit formula for $h(p)$ and was able to calculate $h(p)$ for small values of p.

One of the properties of p needed by Kummer in connection with Fermat's last theorem was the following: p does not divide the class number $h(p)$. Today, a prime with this property is called a *regular prime*.

The second property mentioned by Kummer concerned units, and he showed later that it is satisfied by all regular primes. This was another beautiful result of Kummer, today called Kummer's lemma on units.

In his regularity criterion, Kummer established that the prime p is regular if and only if p does not divide the numerators of the Bernoulli numbers $B_2, B_4, B_6, \ldots, B_{p-3}$ (already defined in Chapter 4, Section I, A).

Kummer was soon able to determine all irregular primes less than 163, namely, 37, 59, 67, 101, 103, 131, 149, 157. He maintained the hope that there exist infinitely many regular primes. This is a truly difficult problem to settle, even though the answer should be positive, as it is supported by excellent numerical evidence.

In 1964, Siegel proved, under heuristic assumptions about the residues of Bernoulli numbers modulo primes, that the density of regular primes among all the primes is $1/\sqrt{e} \cong 61\%$.

On the other hand, it was somewhat of a surprise when Jensen proved, in 1915, that there exist infinitely many irregular primes. The proof was actually rather easy, involving some arithmetical properties of the Bernoulli numbers.

Let $\pi_{reg}(x)$ be the number of regular primes $p \leq x$ (including $p = 2$),

$$\pi_{ii}(x) = \pi(x) - \pi_{reg}(x).$$

For each irregular prime p, the pair $(p, 2k)$ is called an *irregular pair* if $2 \leq 2k \leq p - 3$ and p divides the numerator of B_{2k}. The *irregularity index* of p, denoted $ii(p)$, is the number of irregular pairs $(p, 2k)$.

For $s \geq 1$ let $\pi_{iis}(x)$ be the number of primes $p \leq x$ such that $ii(p) = s$.

RECORD

The most extensive calculations about regular primes (and the validity of Fermat's last theorem) are due to Buhler, Crandall & Sompolski (1991, still unpublished). They continued the work of Wagstaff (1978) and Tanner & Wagstaff (1987), who extended previous computations done by Johnson (1975). Here are the results with $N = 10^6$:

$\pi(N) = 78497$

$\pi_{reg}(N) = 47627$

$\pi_{ii}(N) = 30870$

$\pi_{ii1}(N) = 23816$ (the smallest irregular prime is 37)

$\pi_{ii2}(N) = 5954$ (the smallest prime with irregularity index 2 is 157)

$\pi_{ii3}(N) = 956$

$\pi_{ii4}(N) = 132$

$\pi_{ii5}(N) = 11$

$\pi_{ii6}(N) = 1$ (this is the prime 527377)

$\pi_{iis}(N) = 0$ for all $s \geq 7$.

p to $N = 150000$; according to Tanner & Wagstaff.

The longest string of consecutive regular primes has 27 primes and begins with 17881.

The longest string of consecutive irregular primes has 11 primes and begins with 8597.

The only irregular pairs which are "successive" $(p, 2k)$, $(p, 2k+2)$, are $p = 491$, $2k = 336$ or $p = 587$, $2k = 90$. There are no triples $(p, 2k)$, $(p, 2k + 2)$, $(p, 2k + 4)$ of irregular pairs.

It is conjectured, but it has never been proved, that there are primes with arbitrarily high irregularity index.

Using a result of Kummer and a criterion of Vandiver (see Ribenboim, 1979), it is now known that Fermat's last theorem is true for every prime exponent up to 10^6.

II. Sophie Germain Primes

I have already encountered the Sophie Germain primes, in Chapter 2, in connection with a criterion of Euler about divisors of Mersenne numbers.

I recall that p is a Sophie Germain prime when $2p + 1$ is also a prime. They were first considered by Sophie Germain and she proved the beautiful theorem:

If p is a Sophie Germain prime, then there are no integers x, y, z, different from 0 and not multiples of p, such that $x^p + y^p = z^p$.

In other words, for Sophie Germain's primes the "first case of Fermat's last theorem" is true. For a more detailed discussion, see my own book (1979).

It is presumed that there are infinitely many Sophie Germain primes. However, the proof would be of the same order of difficulty as that of the existence of infinitely many twin primes.

Now I wish to explain in more detail the relations between the first case of Fermat's last theorem and primes like Sophie Germain primes.

Sophie Germain's theorem was extended by Legendre as follows: if $p > 2$ and $4p + 1$, or $8p + 1$, or $10p + 1$, or $14p + 1$, or $16p + 1$, are primes, then the first case of Fermat's last theorem is true for the exponent p. This was further extended by Dénes, as (1951), and more recently by Fee & Granville (1991):

If p is a prime, m is not a multiple of 3, $m \leq 100$, and $2mp + 1$ is also a prime, then the first case of Fermat's last theorem is true for the exponent p.

Now I indicate estimates for the number of Sophie Germain primes less than any number $x \geq 1$.

More generally, let a, $d \geq 1$, with ad even and $\gcd(a, d) = 1$.

For every $x \geq 1$, let

$$S_{d,a}(x) = \#\{\, p \text{ prime } \mid p \leq x, \ a + pd \text{ is a prime}\,\}.$$

If $a = 1$, $d = 2$, then $S_{2,1}(x)$ counts the Sophie Germain primes $p \leq x$.

The same sieve methods of Brun, used to estimate the number $\pi_2(x)$ of twin primes less than x, yield here the similar bound

$$S_{d,a}(x) < \frac{Cx}{(\log x)^2},$$

where $C > 0$ is a constant.

By the prime number theorem,

$$\lim_{x \to \infty} \frac{S_{d,a}(x)}{\pi(x)} = 0.$$

It is then reasonable to say that the set of primes p such that $a + pd$ is prime has density 0. In particular, the set of Sophie Germain primes, and by the same token, the set of twin primes, have density 0.

In 1980, Powell gave a proof avoiding the use of the above facts, sieve methods.

Many large Sophie Germain primes have now been found.

RECORD

The largest Sophie Germain prime known today has been discovered in 1986 by W. Keller: $39051 \times 2^{6001} - 1$. Other large known Sophie Germain primes are $296385 \times 2^{4251} - 1$, $53375 \times 2^{4203} - 1$ discovered by J. Brown, L.C. Noll, B. Parady, G. Smith, J. Smith, S. Zarantonello.

A topic closely related to Sophie Germain primes is the following:

A sequence $q_1 < q_2 < \cdots < q_k$ of primes is a *Cunningham chain of primes of the first kind* (respectively *of second kind*) if each q_i is a prime and $q_{i+1} = 2q_i + 1$ (respectively $q_{i+1} = 2q_i - 1$) for every $i = 1, 2, \ldots, k - 1$. The length of the chain is k. Clearly, the primes in Cunningham chains of the first kind are Sophie Germain primes.

It is not known whether for every k there is a Cunningham chain (of either kind) of length k.

RECORD

The largest known chain of Sophie Germain primes has length 12 and smallest prime 554688278429. The largest known Cunningham chain of primes of second kind has length 13 and smallest prime 758083947856951. They were discovered by Löh (1989).

III. Wieferich Primes

A prime p satisfying the congruence

$$2^{p-1} \equiv 1 \pmod{p^2}$$

ought to be called a *Wieferich prime*.

Indeed, it was Wieferich who proved in 1909 the difficult theorem:

If the first case of Fermat's last theorem is false for the exponent p, then p satisfies the congruence indicated.

It should be noted that, contrary to the congruence $2^{p-1} \equiv 1$

(mod p) which is satisfied by every odd prime, the Wieferich congruence is very rarely satisfied.

Before the computer age(!), Meissner discovered in 1913, and Beeger in 1922, that the primes $p = 1093$ and 3511 satisfy Wieferich's congruence. If you have not been a passive reader, you must have already calculated that $2^{1092} \equiv 1 \pmod{1093^2}$, in Chapter 2, Section III. It is just as easy to show that 3511 has the same property.

RECORD

Lehmer has shown in 1981 that, with the exceptions of 1093 and 3511, there are no primes $p < 6 \times 10^9$, satisfying Wieferich's congruence.

According to results already quoted in Chapter 2, Sections III and IV, the computations of Lehmer say that the only possible factors p^2 (where p is a prime less than 6×10^9) of any pseudoprime, must be 1093 or 3511. This is confirmed by the numerical calculations of Pomerance, Selfridge & Wagstaff, quoted at the end of Chapter 4, Section VI.

In 1910, Mirimanoff proved the following theorem, analogous to Wieferich's theorem:

If the first case of Fermat's last theorem is false for the prime exponent p, then $3^{p-1} \equiv 1 \pmod{p^2}$.

It may be verified that 1093 and 3511 do not satisfy Mirimanoff's congruence. These two results, together with the computations of Lehmer, ensure that the first case of Fermat's last theorem is true for every exponent having a prime factor less than 6×10^9. But this is not the record for the first case of Fermat's last theorem. Wait and see how much bigger it is!

More generally, for any base $a \geq 2$ (where a may be prime or composite) one may consider the primes p such that p does not divide a and $a^{p-1} \equiv 1 \pmod{p^2}$. In fact, it was Abel who first asked for such examples (in 1828); these were provided by Jacobi, who indicated the following congruences, with $p \leq 37$:

$$3^{10} \equiv 1 \pmod{11^2}$$
$$9^{10} \equiv 1 \pmod{11^2}$$

$$14^{28} \equiv 1 \pmod{29^2}$$
$$18^{36} \equiv 1 \pmod{37^2}.$$

I will give soon a more extensive table.

The quotient

$$q_p(a) = \frac{a^{p-1} - 1}{p}$$

has been called the *Fermat quotient* of p, *with base* a. The residue modulo p of the Fermat quotient behaves somehow like a logarithm (this was observed already by Eisenstein in 1850): if p does not divide ab, then

$$q_p(ab) \equiv q_p(a) + q_p(b) \pmod{p}.$$

Also

$$q_p(p - 1) \equiv 1 \pmod{p}, \qquad q_p(p + 1) \equiv -1 \pmod{p}.$$

In my article "1093" (1983), I indicated many truly interesting properties of the Fermat quotient.

As an illustration, I quote the following congruence, which is due to Eisenstein (1850):

$$q_p(2) \equiv \frac{1}{p}\left(1 - \frac{1}{2} + \frac{1}{3} - \frac{1}{4} + \cdots - \frac{1}{p-1}\right) \pmod{p}.$$

The following problems are open:

(1) Given $a \geq 2$, do there exist infinitely many primes p such that

$$a^{p-1} \equiv 1 \pmod{p^2}?$$

(2) Given $a \geq 2$, do there exist infinitely many primes p such that

$$a^{p-1} \not\equiv 1 \pmod{p^2}?$$

The answer to (1) should be positive, why not? But I'm stating it with no base whatsoever, since the question is no doubt very difficult. The next question concerns a fixed prime and a variable base:

(3) If p is an odd prime, does there exist one or more bases a, $2 \leq a < p$, such that $a^{p-1} \equiv 1 \pmod{p^2}$?

Few results are known.

Kruyswijk showed in 1966 that there exists a constant C such that for every odd prime p:

$$\#\{\, a \mid 2 \le a < p,\, a^{p-1} \equiv 1 \pmod{p^2}\,\} < p^{\frac{1}{2}+\frac{C}{(\log\log p)}}.$$

So, not too many bases are good for each prime p.

In 1987, Granville proved

$$\#\{\, q \text{ prime} \mid 2 \le q \le p,\ q^{p-1} \equiv 1 \pmod{p^2}\,\} < p^{1/2}$$

and

$$\#\{\, q \text{ prime} \mid 2 \le q \le p,\ q^{p-1} \not\equiv 1 \pmod{p}\,\} \ge \pi(p) - p^{1/2}.$$

Powell showed that if $p \not\equiv 7 \pmod 8$, then there is at least one prime $q < \sqrt{p}$ such that $q^{p-1} \not\equiv 1 \pmod{p^2}$ (problem posed in 1982 to the *American Mathematical Monthly*, solution published in 1986, by Tzanakis). In this respect, the best result—is by Granville (1987). He showed that if u is any positive integer and p any prime such that $p \ge u^{2u}$ then

$$\#\{\, q \text{ prime} \mid 2 \le q \le u^{1/u}, q^{p-1} \equiv 1 \pmod{p^2}\,\} < u p^{1/2u}.$$

RECORD

Besides the computations of Lehmer for the Fermat quotient with base 2, Brillhart, Tonascia & Weinberger considered (in 1971) bases up to 99. Below is an extension of the table by Kloss (1965) for odd prime bases up to 97:

Base	Range up to	Prime solutions p of $a^{p-1} \equiv 1 \pmod{p^2}$
3	2^{30}	11 1006003
5	2^{29}	20771 40487 53471161
7	2^{28}	5 491531
11	2^{28}	71
13	2^{28}	863 1747591
17	2×10^8	3 46021 48947
19	2×10^8	3 7 13 43 137 63061489
23	2×10^8	13 2481757 13703077
29	2^{28}	None
31	2×10^8	7 79 6451 2806861*
37	2×10^8	3 77867
41	2×10^8	29 1025273 138200401*
43	2×10^8	5 103
47	2×10^8	None
53	2×10^8	3 47 59 97
59	2×10^8	2777
61	2×10^8	None
67	2×10^8	7 47 268573
71	2×10^8	3 47 331
73	2×10^8	3
79	2×10^8	7 263 3037 1012573* 60312841*
83	2×10^8	4871 13691
89	2×10^8	3 13
97	2×10^8	7 2914393*

The solutions marked with an asterisk have been obtained by Keller, and kindly communicated to me.

The work of Wieferich and Mirimanoff was continued by Frobenius, Vandiver, Pollaczeh Morishima, Rosser; this is described in my book on Fermat's last theorem, already quoted. More recently, the results and calculations were pushed much further, by Granville & Monagan (in 1988) for $\ell \leq 89$, and Suzuki (to appear in 1991) for $\ell = 97, 101, 103$:

If the first case of Fermat's last theorem is false for the exponent p, then every prime $\ell \leq 103$ satisfies the congruence

$$\ell^{p-1} \equiv \pmod{p^2}.$$

The combination of these congruences implies the validity of the first case of Fermat's last theorem for very large exponents; the method was devised by Gunderson (1948) and refined by Tanner & Wagstaff (1989), and Coppersmith (1990).

RECORD

The first case of Fermat's last theorem is true for every prime exponent

$$p \leq 2.327 \times 10^{19}.$$

The first case of Fermat's last theorem holds also for much larger primes, in particular, for every Mersenne prime—this follows, as noted by Landau, easily from the congruences of Wieferich and Mirimanoff.

IV. Wilson Primes

This is a very short section—almost nothing is known.

Wilson's theorem states that if p is a prime, then $(p-1)! \equiv -1$ (mod p), thus the so-called *Wilson quotient*

$$W(p) = \frac{(p-1)! + 1}{p}$$

is an integer.

p is called a *Wilson prime* when $W(p) \equiv 0$ (mod p), or equivalently $(p-1)! \equiv -1$ (mod p^2).

For example, $p = 5, 13$ are Wilson primes.

It is not known whether there are infinitely many Wilson primes. In this respect, Vandiver wrote:

> This question seems to be of such a character that if I should come to life any time after my death and some mathematician were to tell me it had definitely been settled, I think I would immediately drop dead again.

RECORD

There is only one other known Wilson prime, besides 5, 13. It is 563, which was discovered by Goldberg in 1953 (one of the first successful computer searches, involving very large numbers).

The search for Wilson primes was continued by E.H. Pearson, K.E. Kloss, W. Keller, H. Dubner and finally by R.H. Gonter &

E.G. Kundert, in 1988, up to 10^7; no new Wilson prime has been found.

V. Repunits and Similar Numbers

There has been a great curiosity about numbers all of whose digits (in the base 10) are equal to 1: 1, 11, 111, 1111, ... They are called *repunits*. When are such numbers prime?

The notation Rn is commonly used for the number

$$11\ldots1 = \frac{10^n - 1}{9},$$

with n digits equal to 1. If Rn is a prime, then n must be a prime, because if $n, m > 1$ then

$$\frac{10^{nm} - 1}{9} = \frac{10^{nm} - 1}{10^m - 1} \times \frac{10^m - 1}{9}$$

and both factors are greater than 1.

RECORD

The only known prime repunits are R2, R19, R23, and more recently R317 (discovered by Williams in 1978) and R1031 (discovered by Williams & Dubner in 1986). Moreover, Dubner has determined that for every other prime p less than 10000, the repunit Rp is composite. Actual factorizations are now known for every $p \leq 89$.

Open problem: are there infinitely many primes repunits?

For more information about repunits see, for example, the book by Yates (1982).

A repunit different from 1 cannot be a square (very easy to see), cannot be a cube (more delicate to show; Rotkiewicz, 1987), cannot be a fifth power (kindly communicated to me by Bond and also by Inkeri, 1989). It is not known if such repunits can be a kth power, for any k not a multiple of 2, 3 or 5; see Obláth, (1956).

Williams & Seah considered also in 1979 numbers of the form $(a^n - 1)/(a - 1)$, where $a \neq 2, 10$. It is usually difficult to establish the primality of large numbers of this type. Yet, in the last ten years much progress has been made both in technology and in the theory of factorization and primality testing.

Below is a table of primes of the form $\frac{a^n - 1}{a - 1}$ for $n \leq 1000$, $a = 3,5,6,7,11,12$ and for $n \leq 10000$, $a = 10$ (repunits).

a	n								
3	3	7	13	71	103	541			
5	3	7	11	13	47	127	149	181	619	929
6	2	3	7	29	71	127	271	509	
7	5	13	131	149					
10	2	19	23	317	1031				
11	17	19	73	139	907				
12	2	3	5	19	97	109	317	353	

VI. Numbers $k \times 2^n \pm 1$

As I had said in Chapter 2, the factors of Fermat numbers are of the form $k \times 2^n + 1$. This property brought these numbers into focus, so it became natural to investigate their primality.

Besides the Mersenne numbers, other numbers of the form $k \times 2^n - 1$, as well as numbers of the form $k^2 \times 2^n + 1$, $k^4 \times 2^n + 1$, $k \times 10^n + 1$, have also been tested for primality.

From Dirichlet's theorem on primes in arithmetic progressions, given $n \geq 1$, there exist infinitely many integers $k \geq 1$, such that $k \times 2^n + 1$ is a prime, and also infinitely many integers $k' \geq 1$, such that $k' \times 2^n - 1$ is a prime.

On the other hand, it is interesting to ask the following question: Given $k \geq 1$, does there exist some integer $n \geq 1$, such that $k \times 2^n + 1$ is prime?

This question was asked by Bateman and answered by Erdös & Odlyzko (in 1979). I quote special cases of their results.

For any real number $x \geq 1$, let $N(x)$ denote the number of odd integers k, $1 \leq k \leq x$, such that there exists $n \geq 1$ for which $k \times 2^n + 1$ is a prime. Then there exists $C_1 > 0$, C_1 being effectively computable, such that $N(x) \geq C_1 x$ (for every $x \geq 1$).

The method developed serves also to study the sequences $k \times 2^n - 1$ ($n \geq 1$) and other similar sequences.

Before discussing the records, note that it is not interesting to ask for the largest prime of the form $k \times 2 + 1$; just take $k = 2^{q-1} - 1$, with $q = 216091$, to obtain the largest Mersenne prime. So, n should be greater than 1.

RECORD

The largest prime known today is $391581 \times 2^{216193} - 1$, which supplanted the Mersenne prime M_{216091}. It was discovered by six devoted numerologists (J. Brown, L.C. Noll, B. Parady, G. Smith, J. Smith, S. Zarantonello) in August 1989. The second largest non-Mersenne prime was discovered by the same group: $235235 \times 2^{70000} - 1$.

For the numbers of the form $k \times 2^n + 1$, with $n \geq 2$, the largest ones were discovered by Buell & Young in 1988 and 1987: $8423 \times 2^{59877} + 1$, $8423 \times 2^{55157} + 1$. Not long ago, the record in this contest was held by Keller, Dubner.

The largest known prime of the form $k^2 \times 2^n + 1$ was discovered by Keller in 1984: $17^2 \times 2^{18502} + 1 = (17 \times 2^{9251})^2 + 1$. This is also the largest known prime of the form $n^2 + 1$.

The largest known prime of the form $k^4 \times 2^n + 1$ is $6952^4 \times 2^{9952} + 1$ (Atkin & Rickert, private communication). It is equal to $(869 \times 2^{2491})^4 + 1$ and it is also the largest known prime of the form $n^4 + 1$.

The largest known prime of the form $k \times 10^n + 1$ is $150093 \times 10^{8000} + 1$, with 8006 digits. It was discovered by Dubner on March 12, 1986; it took approximately 60 days of computing (24 hours per day) over a period of 7 months!

A *palindrome number* (in decimal base) is an integer $N = a_1 a_2 \ldots a_n$, with digits a_i $0 \leq a_i \leq 9$, such that $a_1 = a_n$, $a_2 = a_{n-1}$, $a_3 = a_{n-2}$, etc.

These numbers have absolutely no interesting intrinsic properties, somewhat like repunits. Yet, as a remnant of the old mysticism involving numbers (e.g. perfect, abundant numbers) they still attract the attention of numerologists.

In 1989, Dubner found the two largest known palindromic primes, namely:

$$(10^{5004} + 1232321) \times 10^{4998} + 1 = 1_{0_{4997}}1232321_{0_{4997}}1$$

(this notation means: the digit 1, followed by 4947 digits equal to 1, then in succession the digits 1232321, etc.); this is the largest one, having 10003 digits.

$$(10^{4507} + 2166612) \times 10^{4501} + 1 = 1_{0_{4500}}2166612_{0_{4500}}1,$$

and it has 9009 digits.

Even though a positive proportion of integers $k \geq 1$ has the property that $k \times 2^n + 1$ is a prime (for some n), Sierpiński proved in 1960 the following interesting theorem:

There exist infinitely many odd integers k such that $k \times 2^n + 1$ is composite (for every $n \geq 1$).

The numbers k with the above property are called *Sierpiński numbers*.

RECORD

The smallest known Sierpiński number is $k = 78557$ (Selfridge, 1963).

In 1983, superseding work done by Jaeschke, Keller showed that if there exists a smaller Sierpiński number, it must be one of a list of 69 odd numbers. In a letter of September 1986, Keller writes that he has now reduced further this list to only 57 odd numbers.

It follows from Dirichlet's theorem on primes in arithmetic progressions and Sierpiński's result that there exists infinitely many Sierpiński numbers which are primes.

ADDENDUM ON CULLEN NUMBERS

The numbers of the form $Cn = n \times 2^n + 1$ are known as *Cullen numbers*. In 1958, Robinson showed the C141 is a prime, the only known Cullen prime for 25 years, apart from C1 = 3. In 1984, Keller showed that Cn is also prime for $n = 4713, 5795, 6611, 18496$, and for every other $n \leq 20000$, Cn is composite (private communication).

The numbers $C'n = n \times 2^n - 1$ have also been called Cullen numbers. Of these, the largest primes are for $n = 5312, 7755, 9531, 12379, 15822$. They were discovered by Keller (the last two in 1987).

VII. Primes and Second-Order Linear Recurrence Sequences

In this short section I shall consider sequences $T = (T_n)_{n \geq 0}$, defined by second-order linear recurrences.

Let P, Q be given nonzero integers such that $D = P^2 - 4Q \neq 0$. These integers P, Q are the parameters of the sequence T, to be defined now.

Let T_0, T_1 be arbitrary integers and for every $n \geq 2$, let

$$T_n = PT_{n-1} - QT_{n-2}.$$

The characteristic polynomial of the sequence T is $X^2 - PX + Q$; its roots are

$$\alpha = \frac{P + \sqrt{D}}{2}, \quad \beta = \frac{P - \sqrt{D}}{2}.$$

So $\alpha + \beta = P$, $\alpha\beta = Q$, $\alpha - \beta = \sqrt{D}$.

The sequences $(U_n)_{n \geq 0}$, $(V_n)_{n \geq 0}$, with parameters (P, Q) and $U_0 = 0$, $U_1 = 1$, resp. $V_0 = 2$, $V_1 = P$, are precisely the Lucas sequences, already considered in Chapter 2, Section IV.

Let $\gamma = T_1 - T_0\beta$, $\delta = T_1 - T_0\alpha$, then

$$T_n = \frac{\gamma\alpha^n - \delta\beta^n}{\alpha - \beta} = T_1 \frac{\alpha^n - \beta^n}{\alpha - \beta} - QT_0 \frac{\alpha^{n-1} - \beta^{n-1}}{\alpha - \beta},$$

for every $n \geq 0$.

If $U = (U_n)_{n \geq 0}$ is the Lucas sequence with the same parameters, then $T_n = T_1 U_n - QT_0 U_{n-1}$ (for $n \geq 2$).

It is also possible to define the companion sequence $W = (W_n)_{n \geq 0}$. Let

$$W_0 = 2T_1 - P_0 T, \qquad W_1 = T_1 P - 2QT_0$$

and

$$W_n = PW_{n-1} - QW_{n-2}, \qquad \text{for } n \geq 2.$$

Again, $W_n = \gamma\alpha^n + \delta\beta^n = T_1 V_n - QT_0 V_{n-1}$, where $V = (V_n)_{n \geq 0}$ is the companion Lucas sequence with parameters (P, Q).

Now, I could proceed establishing many algebraic relations and divisibility properties of the sequences, just like for the Lucas sequences, in Chapter 2, Section IV. However, my aim is just to discuss properties related with prime numbers.

First, consider the set

$$\mathcal{P}(T) = \{ p \text{ prime} \mid \text{there exists } n \text{ such that } p | T_n \}.$$

The sequence T is called *degenerate* if $\alpha/\beta = \eta$ is a root of unity. Then $\beta/\alpha = \eta^{-1}$ is also a root of unity; thus $|\eta + \eta^{-1}| \leq 2$. But

$$\eta + \eta^{-1} = \frac{\alpha_+^2 \beta^2}{\alpha\beta} = \frac{P^2 - 2Q}{Q},$$

so, if T is degenerate then $P^2 - 2Q = 0$, $\pm Q$, $\pm 2Q$.

It is not difficult to show that if T is degenerate, then $\mathcal{P}(T)$ is finite.

In 1954, Ward proved that the converse is true:

If T is any nondegenerate sequence, then $\mathcal{P}(T)$ is infinite.

A natural problem is to ask whether $\mathcal{P}(T)$ has necessarily a positive density (in the set of all primes) and, if possible, to compute it.

The pioneering work was done by Hasse (1966). He wanted to study the set of primes p such that the order of 2 modulo p is even. This means that there exists $n \geq 1$ such that p divides $2^{2n} - 1$, but p does not divide $2^m - 1$, for $1 \leq m < 2n$. Hence $2^n \equiv -1 \pmod{p}$, so p divides $2^n + 1$, and conversely.

The sequence $H = (H_n)_{n \geq 0}$, with $H_n = 2^n + 1$, is the companion Lucas sequence with parameters $P = 3$, $Q = 2$.

Let $\pi_H(x) = \#\{\, p \in \mathcal{P}(H) \mid p \leq x \,\}$ for every $x \geq 1$.

Hasse showed that

$$\lim_{x \to \infty} \frac{\pi_H(x)}{\pi(x)} = \frac{17}{24}.$$

The number $17/24$ represents the density of primes dividing the sequence H.

However, if $Q \geq 3$ and Q is square-free, if $H' = (H'_n)_{n \geq 0}$ is the companion Lucas sequence with parameters $(Q + 1, Q)$, then $H'_n = Q^n + 1$. For this sequence, the density of $\mathcal{P}(H')$ is $\frac{2}{3}$.

In 1985, Lagarias reworked Hasse's method and showed, among other results, that for the sequence $V = (V_n)_{n \geq 0}$ of Lucas numbers, $\mathcal{P}(V)$ has density $\frac{2}{3}$.

The prevailing conjecture is that for every nondegenerate sequence T, the set $\mathcal{P}(T)$ has a positive density.

Now, I turn to another very interesting and difficult problem.

Let $T = (T_n)_{n \geq 0}$ be a second-order linear recurring sequence. For example, the Fibonacci or Lucas sequences. These sequences do contain prime numbers, but it is not known, and certainly difficult to decide, whether they contain infinitely many prime numbers.

It follows from Chapter 2, Section IV, relations (IV.15), resp. (IV.16):

If U_m is a prime (and $m \neq 4$) then m is a prime.
If V_m is a prime (and m is not a power of 2) then m is a prime.

Of course, the converse need not be true.

For $3 \leq n < 1000$, the only Fibonacci numbers which are primes are the U_n, with:

$$n = 3, 4, 5, 7, 11, 13, 17, 23, 29, 43, 47, 83, 131, 137, 359,$$
$$431, 433, 449, 509, 569, 571.$$

The above list was indicated by Brillhart (1963). To my knowledge, the largest Fibonacci number recognized to be a prime, is U_{2971}, discovered by Williams.

Concerning the Lucas numbers, if $0 \leq n \leq 500$, L_n is a prime exactly when

$$n = 0, 2, 4, 5, 7, 8, 11, 13, 16, 17, 19, 31, 37, 41, 47, 53, 61,$$
$$71, 79, 113, 313, 353$$

(as listed by Brillhart). More prime Lucas numbers, discovered by Williams, are V_{503}, V_{613}, V_{617}, V_{863}, the latter being, according to the information at my disposal, the largest known prime Lucas numbers.

It should be noted that, if T is neither a Lucas sequence nor a companion Lucas sequence, then T need not to contain any prime number. The following example was found by Graham in 1964: let $P = 1$, $Q = -1$ and

$$T_0 = 1786772701928802632268715130455793,$$
$$T_1 = 1059683225053915111058165141686995.$$

Then T contains no primes.

6

Heuristic and Probabilistic Results about Prime Numbers

The word "heuristic" means: based on, or involving, trial and error. Heuristic results are formulated following the observation of numerical data from tables or from extended calculations. Sometimes these results express the conclusions of some statistical analysis.

There are also probabilistic methods. The idea is quite well explained in Cramér's paper (1937) already quoted in Chapter 4:

> In investigations concerning the asymptotic properties of arithmetic functions, it is often possible to make an interesting use of probability arguments. If, e.g., we are interested in the distribution of a given sequence S of integers, we then consider S as a member of an infinite class C of sequences, which may be concretely interpreted as the possible realizations of some game of chance. It is then in many cases possible to prove that, *with a probability equal to* 1, a certain relation R holds in C, i.e., that in a definite mathematical sense, "almost all" sequences of C satisfy R. Of course we cannot, in general, conclude that R holds for the particular sequence S, but results suggested in this way may sometimes afterwards be rigorously proved by other methods.

Heuristic and probabilistic methods, if not properly handled with caution and intelligence, may give rise to "dream mathematics," far removed from the reality. Hasty conjectures, misinterpretation of numerical evidence have to be avoided.

I'll be careful and restrict myself to only a few of the reliable contributions. I have in mind, Hardy & Littlewood, with their famous conjectures in *Partitio Numerorum*. Dickson, Bouniakowsky, and Schinzel & Sierpiński, with their intriguing hypotheses.

I. Prime Values of Linear Polynomials

Once, more, the starting point is Dirichlet's theorem on primes in arithmetic progressions. It states that if $f(X) = bX+a$, with integers a, b such that $a \neq 0$, $b \geq 1$, $\gcd(a,b) = 1$, then there exist infinitely many integers $m \geq 0$ such that $f(m)$ is a prime.

In 1904, Dickson stated the following conjecture, concerning the simultaneous values of several linear polynomials:

(D) *Let $s \geq 1$, $f_i(X) = b_iX + a_i$ with a_i, b_i integers, $b_i \geq 1$ (for $i = 1,\ldots,s$). Assume that the following condition is satisfied:*

(*) There does not exist any integer $n > 1$ dividing all the products $f_1(k)f_2(k)\cdots f_s(k)$, for every integer k.

Then there exist infinitely many natural numbers m such that all numbers $f_1(m), f_2(m),\ldots,f_s(m)$ are primes.

The following statement looks weaker than (D):

(D_0) *Under the same assumptions for $f_1(X),\ldots,f_s(X)$, there exists a natural number m such that the numbers $f_1(m),\ldots,f_s(m)$ are primes.*

Whereas, at first sight, one might doubt the validity of (D), the statement (D_0) is apparently demanding so much less, that it may be more readily accepted. But, the fact is that (D) and (D_0) are equivalent.

Indeed, if (D_0) is true, there exists $m_1 \geq 0$ such that $f_1(m_1)$, ..., $f_s(m_1)$ are primes. Let $g_i(X) = f_i(X+1+m_1)$ for $i = 1,\ldots,s$. Then (*) is satisfied by $g_1(X),\ldots,g_s(X)$, hence by (D_0) there exists $k_1 \geq 0$ such that $g_1(k_1),\ldots,g_s(k_1)$ are primes; let $m_2 = k_1+1+m_1 > m_1$, so $f_1(m_2),\ldots,f_s(m_2)$ are primes. The argument may be repeated, and shows that (D_0) implies (D).

Dickson did not explore the consequences of his conjecture. This was the object of a paper by Schinzel & Sierpiński (1958), which I find so interesting!

As a matter of fact, Schinzel had proposed a more embracing conjecture [the hypothesis (H)] dealing with polynomials not necessarily linear. But, before I discuss the hypothesis (H) and its consequences, I shall indicate the many interesting results which Schinzel & Sierpiński proved, under the assumption that the conjecture (D) is valid.

The impressive consequences of the hypothesis (D), listed below, should convince anyone that the proof of (D) is remote, if ever possible.

(D_1) *Let $s \geq 1$, let $a_1 < a_2 < \cdots a_s$ be nonzero integers and assume that $f_1(X) = X + a_1, \ldots, f_s(X) = X + a_s$ satisfy condition (*) in (D).*

Then there exist infinitely many integers $m \geq 1$ such that $m + a_1$, $m + a_2$, $\ldots$, $m + a_s$ are consecutive primes.

The conjecture of Polignac (1849), discussed in Chapter 4, Sections II and III is a consequence of (D_1):

($D_{1,1}$) *For every even integer $2k \geq 2$ there exist infinitely many pairs of consecutive primes with difference equal to $2k$. In particular, there exist infinitely many pairs of twin primes.*

Here is an interesting consequence concerning the abundance of twin primes:

($D_{1,2}$) *For every integer $m \geq 1$, there exist $2m$ consecutive primes which are m couples of twin primes.*

Another quite unexpected consequence concerns primes in arithmetic progressions. In Chapter 4, Section IV, I showed that if a, $a+d$, $\ldots$, $a+(n-1)d$ are primes and $1 < n < a$ then d is a multiple of $\prod_{p \leq n} p$.
From (D_1), it follows:

($D_{1,3}$) *Let $1 < n$, let d be a multiple of $\prod_{p \leq n} p$. Then there exist infinitely many arithmetic progressions, with difference d, each consisting of n consecutive primes.*

The reader should compare this very strong statement with what was stated independently of any conjecture in Chapter 4, Section IV.
Concerning Sophie Germain primes, it can be deduced from (D):

(D_2) *For every $m \geq 1$ there exists infinitely many arithmetic progressions consisting of m Sophie Germain primes.*

In particular, (D) implies the existence of infinitely many Sophie Germain primes, a fact which has never been proved without appealing to a conjecture. I shall return soon to a quantitative statement about the distribution of Sophie Germain primes.
The conjecture (D) is so powerful that it implies also:

(D_3) *There exist infinitely many composite Mersenne numbers.*

I recall (see Chapter 2, Section IV) that no one has as yet succeeded in proving [without assuming the conjecture (D)] that there exist infinitely many composite Mersenne numbers. However, it is easy to prove that other sequences, similar to that of Mersenne numbers, contain infinitely composite numbers. The following result was proposed by Powell as a problem in 1981 (a solution by Israel was published in 1982):

If m, n are integers such that $m > 1$, $mn > 2$ (this excludes $m = 2$, $n = 1$), then there exist infinitely many composite numbers of the form $m^p - n$, where p is a prime number.

Proof. Let q be a prime dividing $mn - 1$, so $q \nmid m$. If p is a prime such that $p \equiv q - 2 \pmod{q - 1}$, then $m(m^p - n) \equiv m(m^{q-2} - n) \equiv 1 - mn \equiv 0 \pmod{q}$, hence q divides $m^p - n$. By Dirichlet's theorem on primes in arithmetic progressions, there exist infinitely many primes p, such that $p \equiv q - 2 \pmod{q - 1}$, hence there exist infinitely many composite integers $m^p - n$, where p is a prime number. □

Another witness of the strength of conjecture (D) is the following:

(D_4) *There exist infinitely many composite integers n, such that for every a, $1 < a < n$, then $a^n \equiv a \pmod{n}$ (so n is an a-pseudoprime, for every such a). In particular, there exist infinitely many Carmichael numbers.*

II. Prime Values of Polynomials of Arbitrary Degree

Now I turn to polynomials which may be nonlinear.

Historically, the first conjecture was by Bouniakowsky, in 1857, and it concerns one polynomial of degree at least two:

(B) *Let $f(X)$ be an irreducible polynomial, with integral coefficients, positive leading coefficient and degree at least 2. Assume that the following condition is satisfied:*

(*) there does not exist any integer $n > 1$ dividing all the values $f(k)$, for every integer k.

Then there exist infinitely many natural numbers m such that $f(m)$ is a prime.

Just as for the conjectures (D) and (D_0), the following conjecture is equivalent to (B):

(B_0) *If $f(X)$ satisfies the same hypothesis as in (B) there exists a natural number m such that $f(m)$ is a prime.*

Before I discuss conjecture (B), let it be made clear that there are, in fact, very few results about prime values of polynomials. To wit, it has never been found even one polynomial $f(X)$ of degree greater than 1, such that $|f(n)|$ is a prime, for infinitely many natural numbers n.

On the other hand, Sierpiński showed in 1964, that for every $k \geq 1$ there exists an integer b such that $n^2 + b$ is a prime for at least k numbers n.

If $f(X)$ has degree $d \geq 2$ and integral coefficients, for every $x \geq 1$ let

$$\pi_{f(X)}(x) = \#\{\, n \geq 1 \mid |f(n)| \leq x \text{ and } |f(n)| \text{ is a prime}\,\}.$$

In 1922, Nagell showed that $\lim_{x \to \infty} \pi_{f(X)}(x)/x = 0$, so there are few prime values. In 1931, Heilbronn proved the more precise statement:

There exists a positive constant C (depending on $f(X)$), such that

$$\pi_{f(X)}(x) \leq C \frac{x}{\log x}, \qquad \text{for every } x \geq 1.$$

Not much more seems to be known for arbitrary polynomials. However, for special types of polynomials, there are interesting conjectures and extensive computations. I'll soon discuss this in more detail.

The question of existence of infinitely many primes p such that $f(p)$ is a prime, is even more difficult. In particular, as I have already said, it is not known that there exist infinitely many primes for which the polynomials $f(X) = X + 2$, respectively $f(X) = 2X + 1$, assume prime values (existence of infinitely many twin primes, or infinitely many Sophie Germain primes). However, if happiness comes from almost-primes, then there are good reasons to be contented, once more with sieve methods and the unavoidable book of Halberstam & Richert. The latter author proved in 1969:

Let $f(X)$ be a polynomial with integral coefficients, positive leading coefficient, degree $d \geq 1$ (and different from X). Assume that for every prime p, the number $\rho(p)$ of solutions of $f(X) \equiv 0 \pmod{p}$ is less than p; moreover if $p \leq d+1$ and p does not divide $f(0)$ assume also that $\rho(p) < p - 1$. Then, there exist infinitely many primes p such that $f(p)$ is a $(2d + 1)$-almost prime.

The following particular case was proved by Rieger (1969): there exist infinitely many primes p such that $p^2 - 2 \in P_5$.

Now, back to Bouniakowsky's hypothesis!

Bouniakowsky drew no consequences from his conjecture. Once more, this was the task of Schinzel & Sierpiński, who studied an even more general hypothesis.

The following proposition, which has never been proved, follows easily if the conjecture (B) is assumed to be true:

(B_1) *Let a, b, c be relatively prime integers, such that $a \geq 1$ and $a + b$ and c are not simultaneously even.*

If $b^2 - 4ac$ is not a square, then there exist infinitely many natural numbers m such that $am^2 + bm + c$ is a prime number.

(B_1) in turn implies:

(B_2) *If k is an integer such that $-k$ is not a square, there exist infinitely many natural numbers m such that $m^2 + k$ is a prime number.*

In particular, (B) implies that there exist infinitely many primes of the form $m^2 + 1$.

The "game of primes of the form $m^2 + 1$" is not just as innocent as one would inadvertently think at first. It is deeply related with the class number of real quadratic fields, but I have to refrain from discussing this matter.

The following statement was also conjectured by Hardy & Littlewood, in 1923; it may be proved assuming the truth of (B).

(B_3) *Let d be an odd integer, $d > 1$; let k be an integer which is not a eth power of an integer, for any factor $e > 1$ of d. Then there exist infinitely many natural numbers m such that $m^d + k$ is a prime.*

In his joint paper with Sierpiński, Schinzel proposed the following conjectures:

(H) *Let $s \geq 1$, let $f_1(X), \ldots, f_s(X)$ be irreducible polynomials, with integral coefficients and positive leading coefficient. Assume that the following condition holds:*

(*) there does not exist any integer $n > 1$ dividing all the products $f_1(k)f_2(k)\ldots f_s(k)$, for every integer k.

Then there exist infinitely many natural numbers m such that all numbers $f_1(m), f_2(m), \ldots, f_s(m)$ are primes.

(H_0) *Under the same assumptions for $f_1(X), \ldots, f_s(X)$, there exists a natural number m such that the numbers $f_1(m), \ldots, f_s(m)$ are primes.*

Once again, (H) and (H_0) are equivalent. If all the polynomials $f_1(X), \ldots, f_s(X)$ have degree 1, these conjectures are Dickson's (D), (D_0). If $s = 1$, they are Bouniakowsky's conjectures (B), (B_0).

I shall not enumerate here the various consequences of this hypothesis, which were proved by the authors.

Still in the same paper, there is the following conjecture by Sierpiński:

(S) *For every integer $n > 1$, let the n^2 integers $1, 2, \ldots, n^2$ be written in an array with n rows, each with n integers, like an $n \times n$ matrix:*

1	2	$\cdots$	n
$n+1$	$n+2$	$\cdots$	$2n$
$2n+1$	$2n+2$	$\cdots$	$3n$
$\vdots$	$\vdots$	$\ddots$	$\vdots$
$(n-1)n+1$	$(n-1)n+2$	$\cdots$	n^2.

Then, there exists a prime number in each row.

Of course, 2 is in the first row. By the theorem of Bertrand and Tschebycheff, there is a prime number in the second row.

More can be said about the first few rows using a strengthening of Bertrand and Tschebycheff's theorem. For examples, in 1932, Breusch showed that if $n \geq 48$, then there exists a prime between n and $(9/8)n$. Thus, if $0 < k \leq 7$ and $n \geq 48$, then there exists a prime p such that $kn + 1 \leq p \leq (9/8)(kn + 1) \leq (k + 1)n$. By direct calculation, the same result also holds for $n \geq 9$. So, there is a prime in each of the first 8 rows.

By the prime number theorem, for every $h \geq 1$ there exists $n_0 = n_0(h) > h$ such that, if $n \geq n_0$, then there exists a prime p such that $n < p < (1 + \frac{1}{h})n$.

And from this, it follows that, if $n \geq n_0$, then each of the first h rows of the array contains a prime.

Just like the preceding conjectures, (S) has also several interesting consequences.

(S_1) *For every $n \geq 2$ there exist at least two primes p, p' such that $(n-1)^2 < p < p' < n^2$.*

($S_{1,1}$) *For every $n \geq 1$ there exist at least four primes p, p', p'', p''' such that $n^3 < p < p' < p'' < p''' < (n+1)^3$.*

It should be noted that both statements (S_1), ($S_{1,1}$) have not yet been proved without appealing to the conjecture (S). However, it is easy to show, that for every sufficiently large n, there exists a prime p between n^3 and $(n+1)^3$; this is done using Ingham's result that $d_n = O(p_n^{(5/8)+\varepsilon})$ for every $\varepsilon > 0$.

Schinzel stated the following "transposed" form of Sierpiński's conjecture:

(S') *For every integer $n > 1$, let the n^2 integers $1, 2, \ldots, n^2$ be written in an array with n rows, each with n integers [just like in (S)]. If $1 \leq k \leq n$ and $\gcd(k, n) = 1$, then the kth column contains at least one prime number.*

This time, Schinzel & Sierpiński drew no consequence from the conjecture (S'). I suppose it was Sunday evening and they were tired. However, in 1963, this conjecture was spelled out once more by Kanold.

I conclude with the translation of the following comment in Schinzel & Sierpiński's paper:

> We do not know what will be the fate of our hypotheses, however we think that, even if they are refuted, this will not be without profit for number theory.

III. Polynomials with Many Successive Composite Values

I want now to report on the new results of McCurley, which I find very interesting. According to Bouniakowsky's conjecture, if $f(X)$ is an irreducible polynomial with integral coefficients and satisfying condition (*), then there exists a smallest integer $m \geq 1$ such that $f(m)$ is a prime. Denote it by $p(f)$.

If $f(X) = dX + a$, with $d \geq 2$, $1 \leq a \leq d - 1$, $\gcd(a, d) = 1$, then of course, $p(f)$ exists. With the notation of Chapter 4, Section IV,

$$p(dX + a) = \frac{p(d, a) - a}{d}.$$

The work of McCurley concerns polynomials $f(X)$ of higher degree and lower bounds for $p(f)$.

In 1984, McCurley proved:

For every $\varepsilon > 0$ there exist infinitely many irreducible polynomials $f_i(X) = X^{d_i} + k_i$ (with $2 \leq d_1 < d_2 < \cdots$, with each $k_i > 1$), such that if

$$m < M_i = e^{(C-\varepsilon)(\log d_i)/(\log \log d_i)},$$

then $f_i(m)$ is composite.

This may be rewritten as $p(f_i) \geq M_i$.

It should be noted that the proof gives no explicit polynomials. However, in a computer search, the following polynomials were discovered (the first example is due to Shanks, 1971):

$f(X)$	$f(m)$ is composite for all m up to
$X^6 + 1091$	3905
$X^6 + 82991$	7979
$X^{12} + 4094$	170624
$X^{12} + 488669$	616979

The smallest prime value of the last polynomial has no less than 70 digits.

With another method, McCurley showed in 1986:

For every $d \geq 1$ there exists an irreducible polynomial $f(X)$, of degree d, satisfying the condition (*) and such that if

$$|m| < e^{C\sqrt{L(f)/(\log L(f))}},$$

then $|f(m)|$ is composite.

In the above statement, $L(f)$ is the length of $f(X) = \sum_{k=0}^{d} a_k X^k$ which is defined by $L(f) = \sum_{k=0}^{d} \| a_k \|$ and $\| a_k \|$ is the number of digits in the binary expansion of $|a_k|$, with $\| 0 \| = 1$.

Note that this last result is applicable to polynomials of any degree, and also the proof yields explicit polynomials with the desired property.

McCurley determined $p(X^d + k)$ for several polynomials. From his tables, I note:

d	$k \leq m$	$\max p(X^d + k)$
2	10^6	$p(X^2 + 576239) = 402$
3	10^6	$p(X^3 + 382108) = 297$
4	150000	$p(X^4 + 72254) = 2505$
5	10^5	$p(X^5 + 89750) = 339$

IV. Partitio Numerorum

It is instructive to browse through the paper *Partitio Numerorum, III: On the expression of a number as a sum of primes*, by Hardy & Littlewood (1923). There one finds a conscious systematic attempt— for the first time in such a scale—to derive heuristic formulas for the distribution of primes satisfying various additional conditions.

I shall present here a selection of the probabilistic conjectures, extracted from Hardy & Littlewood's paper (I will keep their labeling, since it is classical). The first conjecture concerns Goldbach's problem:

Conjecture A: *Every sufficiently large even number $2n$ is the sum of two primes. The asymptotic formula for the number of representations is*

$$r_2(2n) \sim C_2 \frac{2n}{(\log 2n)^2} \prod_{\substack{p>2 \\ p|n}} \frac{p-1}{p-2}$$

with

$$C_2 = \prod_{p>2} \left(1 - \frac{1}{(p-1)^2}\right) = 0.66016\dots.$$

Note that C_2 is the same as the twin primes constant (see Chapter 4, Section III).

The following conjecture deals with (not necessarily consecutive) primes with given difference $2k$, in particular, with twin primes:

Conjecture B: *For every even integer $2k \geq 2$ there exist infinitely many primes p such that $p + 2k$ is also a prime. For $x \geq 1$, let*

$$\pi_{2k}(x) = \#\{\,p \text{ prime} \mid p + 2k \text{ is prime and } p + 2k \leq x\,\}.$$

Then

$$\pi_{2k}(x) \sim 2C_2 \frac{x}{(\log x)^2} \prod_{\substack{p>2 \\ p\mid k}} \frac{p-1}{p-2},$$

where C_2 is the twin primes constant.

In particular, if $2k = 2$ this gives the asymptotic estimate for the twin primes counting function, already indicated in Chapter 4, Section III.

The conjecture E is about primes of the form $m^2 + 1$:

Conjecture E: *There exist infinitely many primes of the form $m^2 + 1$. For $x > 1$ let*

$$\pi_{X^2+1}(x) = \#\{\,p \text{ prime} \mid p \leq x \text{ and } p \text{ is of the form } p = m^2 + 1\,\}.$$

Then

$$\pi_{X^2+1}(x) \sim C \frac{\sqrt{x}}{\log x}$$

where

$$C = \prod_{p \geq 3} \left(1 - \frac{(-1/p)}{p-1}\right) = 1.37281346\dots$$

and

$$(-1/p) = (-1)^{(p-1)/2}$$

is the Legendre symbol.

The values $\pi_{X2+1}(x)$ predicted by the conjecture are in significant agreement with the actual values indicated below (the last two were computed by Wunderlich in 1973):

x	$\pi_{X_2+1}(x)$	Conjectural $\pi_{X^2+1}(x)$	Ratio
10^6	112	99	0.8862
10^8	841	745	0.8858
10^{10}	6656	5962	0.8918
10^{12}	54110	49684	0.9182
10^{14}	456362	425861	0.9332

Here is the place to mention the result of Iwaniec (1978): there exist infinitely integers $m^2 + 1$ which are 2-almost-primes.

Here is still another conjecture:

Conjecture F: *Let $a > 0$, b, c be integers such that $\gcd(a, b, c) = 1$, $b^2 - 4ac$ is not a square and $a + b$, c are not both even. Then there are infinitely many primes of the form $am^2 + bm + c$ [this was statement (B_1)] and the number $\pi_{aX^2+bX+c}(x)$ of primes $am^2 + bm + c$ which are less than x is given asymptotically by*

$$\pi_{aX^2+bX+c}(x) \sim \frac{\varepsilon C}{\sqrt{a}} \frac{\sqrt{x}}{\log x} \prod_{\substack{p>2 \\ p|\gcd(a,b)}} \frac{p}{p-1}$$

where

$$\varepsilon = \begin{cases} 1 & \text{if } a + b \text{ is odd} \\ 2 & \text{if } a + b \text{ is even} \end{cases}$$

$$C = \prod_{\substack{p>2 \\ p \nmid a}} \left(1 - \frac{\left(\frac{b^2-4ac}{p}\right)}{p-1} \right)$$

and $\left(\frac{b^2-4ac}{p}\right)$ denotes the Legendre symbol.

In particular, the conjecture is applicable to primes of the form $m^2 + k$, where $-k$ is not a square.

As it should be expected, the conjectures of Hardy & Littlewood inspired a considerable amount of computation, intended to determine accurately the constants involved in the formulas, and to verify that the predictions fitted well with the observation. As the constants were often given by slowly convergent infinite products, it was imperative to modify these expressions, to make them more easily computable.

For the special case of polynomials $f_A(X) = X^2 + X + A$ (with $A \geq 1$ an integer), the conjecture F states that

$$\pi_A(x) \sim C(A) \frac{2\sqrt{x}}{\log x},$$

where $\pi_A = \pi_{x^2+x+A}$,

$$C(A) = \prod_{p>2} \left(1 - \frac{\left(\frac{1-4A}{p}\right)}{p-2} \right).$$

Shanks had determined (in 1975) that $C(41) = 3.3197732$. Fung & Williams calculated (in 1990) $C(A)$ and $f_A(10^6)$ for various values of A.

RECORD

For all values calculated so far, the maximum of $C(A)$ is

$$C(132874279528931) = 5.0870883$$

and

$$\pi_{X^2+X+132874279528931}(10^6) = 312975.$$

In comparison, for example

$$\pi_{X^2+X+41}(10^6) = 261080 \quad \text{with} \quad C(41) = 3.3197732,$$

$$\pi_{X^2+X+27941}(10^6) = 286128, \quad \text{with} \quad C(27941) = 3.6319998,$$

and the maximum of $\pi_{X^2+X+A}(10^6)$ is, among the values calculated, equal to

$$\begin{aligned}\pi_{X^2+X+21425625701}(10^6) &= 361841, \text{ with} \\ C(21425625701) &= 4.7073044.\end{aligned}$$

Conclusion

Dear Reader, or better, Dear Friend of Numbers:

You, who have come faithfully to this point, may now be thinking about what you read and learned. Maybe you are already investigating some of the numerous problems, working with your computer ... and brain. I hope that this presentation of various topics and facets of the theory of prime numbers has been pleasant and conveyed an approximate picture of the problems being investigated. I hope you perceived that records often express efforts to solve open problems and the calculations involved bring a new light to the questions. But, I hope mostly that you became convinced of the value of the fine theorems produced by so many brilliant minds. Personally, I look with awe to the monumental achievements and with grand bewilderment to all the deep questions which remain unsolved—and for how long?

In writing this book I wanted to produce a work of synthesis, to develop the theory of prime numbers as a discipline where the natural questions are systematically studied.

In the Introduction, I believe, I made clear the reasons for dividing the book into its various parts, which are devoted to the answer of what I consider the imperative questions. This organization is intended to prevent any young students from having the same impression as I had in my early days (it was, alas, long ago ...) that number theory dealt with multiple unrelated problems.

This justifies the general plan but not the choice of details, especially the ones which are missing (we always cry for the dear absent ones ...). It is evident that a choice had to be made. I wanted this book to be small and lightweight, so the hand could hold it—not a bulky brick, which cannot be carried in a pocket, nor read while waiting or riding in a train (in Canada, we both wait for and ride in trains).

I admit readily that the proofs of the most important theorems are absent. They are usually technically involved and long. Their inclusion would shift the attention from the general structure of the theory towards the particular details. The unhappy reader may find satisfaction by consulting the excellent papers and books noted in the ample bibliography.

Bibliography

General References

The books listed below are highly recommended for the quality of contents and presentation. This list is, of course, not exhaustive.

1909 LANDAU, E. *Handbuch der Lehre der Verteilung der Primzahlen.* Teubner, Leipzig, 1909. Reprinted by Chelsea, Bronx, N.Y., 1974.

1927 LANDAU, E. *Vorlesungen über Zahlentheorie* (in 3 volumes). S. Hirzel, Leipzig, 1927. Reprinted by Chelsea, Bronx, N.Y., 1969.

1938 HARDY, G. H. & WRIGHT, E. M. *An Introduction to the Theory of Numbers.* Clarendon Press, Oxford, 1938 (5th edition, 1979).

1952 DAVENPORT, H. *The Higher Arithmetic.* Hutchinson, London, 1952. Reprinted by Dover, New York, 1983.

1953 TROST, E. *Primzahlen.* Birkhäuser, Basel, 1953 (second edition, 1968).

1957 PRACHAR, K. *Primzahlverteilung.* Springer-Verlag, Berlin, 1957 (2nd edition 1978).

1962 SHANKS, D. *Solved and Unsolved Problems in Number Theory.* Spartan, Washington, 1962. 3rd edition by Chelsea, Bronx, N.Y., 1985.

1963 AYOUB, R.G. *An Introduction to the Theory of Numbers.* Amer. Math. Soc., Providence, R.I., 1963.

1964 SIERPIŃSKI, W. *Elementary Theory of Numbers.* Hafner, New York, 1964 (second edition, North- Holland, Amsterdam, 1987).

1975 ELLISON, W.J. & MENDÈS-FRANCE, M. *Les Nombres Premiers.* Hermann, Paris, 1975.

1976 ADAMS, W.W. & GOLDSTEIN, L.J. *Introduction to Number Theory.* Prentice-Hall, Englewood Cliffs, N.J., 1976.

1981 GUY, R.K.   *Unsolved Problems in Number Theory.* Springer-Verlag, New York, 1981 (second edition, 1990).

1982 HUA, L.K.   *Introduction to Number Theory.* Springer-Verlag, New York, 1982.

Chapter 1

1878 KUMMER, E.E.   Neuer elementarer Beweis, dass die Anzahl aller Primzahlen eine unendliche ist. Monatsber. Akad. d. Wiss., Berlin, 1878/9, 777–778.

1890 STIELTJES, T.J.   Sur la théorie des nombres. Étude bibliographique. An. Fac. Sci. Toulouse, 4, 1890, 1–103.

1924 PÓLYA, G. & SZEGÖ, G.   *Aufgaben und Lehrsätze aus der Analysis*, 2 vols. Springer-Verlag, Berlin, 1924 (4th edition, 1970).

1947 BELLMAN, R.   A note on relatively prime sequences. Bull. Amer. Math. Soc., 53, 1947, 778–779.

1955 FÜRSTENBERG, H.   On the infinitude of primes. Amer. Math. Monthly, 62, 1955, p. 353.

1959 GOLOMB, S.W.   A connected topology for the integers. Amer. Math. Monthly, 66, 1959, 663–665.

1963 MULLIN, A.A.   Recursive function theory (A modern look on an Euclidean idea). Bull. Amer. Math. Soc., 69, 1963, p. 737.

1964 EDWARDS, A.W.F.   Infinite coprime sequences. Math. Gazette, 48, 1964, 416–422.

1967 SAMUEL, P.   *Théorie Algébrique des Nombres.* Hermann, Paris, 1967. English translation published by Houghton-Mifflin, Boston, 1970.

1968 COX, C.D. & VAN DER POORTEN, A.J.   On a sequence of prime numbers. J. Austr. Math. Soc., 1968, 8, 571–574.

1972 BORNING, A.   Some results for $k! \pm 1$ and $2 \cdot 3 \cdot 5 \cdots p + 1$. Math. Comp., 26, 1972, 567–570.

1980 TEMPLER, M.   On the primality of $k! + 1$ and $2 \cdot 3 \cdots \cdots p + 1$. Math. Comp., 34, 1980, 303–304.

1980 WASHINGTON, L.C.   The infinitude of primes via commutative algebra. Unpublished manuscript.

1982 BUHLER, J.P., CRANDALL, R.E. & PENK, M.A.   Primes of the form $n! \pm 1$ and $2 \cdot 3 \cdot 5 \cdots p \pm 1$. Math. of Comp.,

38, 1982, 639–643.

1985 ODONI, R.W.K.   On the prime divisors of the sequence $w_{n+1} = 1 + w_1 w_2 \cdots w_n$. J. London Math. Soc., (2), 32, 1985, 1–11.

1987 DUBNER, H.   Factorial and primorial primes. J. Recr. Math., 19, 1987, 197–203.

Chapter 2

1801 GAUSS, C.F.   *Disquisitiones Arithmeticae.* G. Fleischer, Leipzig, 1801. English translation by A.A. Clarke. Yale Univ. Press, New Haven, 1966. Revised English translation by W.C. Waterhouse. Springer-Verlag, New York, 1986.

1844 EISENSTEIN, F.G.   Aufgaben. Journal f.d. reine u. angew. Math., 27, 1844, p. 87. Reprinted in *Mathematische Werke*, Vol. I, p. 112. Chelsea, Bronx, N.Y., 1975.

1852 KUMMER, E.E.   Über die Erganzungssätze zu den allgemeinem Reciprocitätsgesetzen. Journal f.d. reine u. angew. Math., 44, 1852, 93-146. Reprinted in *Collected Papers*, Vol. I, (edited by A. Weil, 485–538). Springer-Verlag, New York, 1975.

1876 LUCAS, E.   Sur la recherche des grands nombres premiers. Assoc. Française p. l'Avanc. des Sciences, 5, 1876, 61–68.

1877 PEPIN, T.   Sur la formule $2^{2^n} + 1$. C.R. Acad. Sci. Paris, 85, 1877, 329–331.

1878 LUCAS, E.   Théorie des fonctions numériques simplement périodiques. Amer. J. Math., 1, 1878, 184–240 and 289–321.

1878 PROTH, F.   Théorèmes sur les nombres premiers. C.R. Acad. Sci. Paris, 85, 1877, 329–331.

1886 BANG, A.S.   Taltheoretiske Underso/gelser. Tidskrift f. Math., ser. 5,4, 1886, 70–80 and 130–137.

1891 LUCAS, E.   *Théorie des Nombres.* Gauthier-Villars, Paris, 1891. Reprinted by A. Blanchard, Paris, 1961.

1892 ZSIGMONDY, K.   Zur Theorie der Potenzreste. Monatsh. f. Math., 3, 1892, 265–284.

1903 MALO, E.   Nombres qui, sans être premiers, vérifient exceptionnellement une congruence de Fermat. L'Interm. des Math., 10, 1903, p. 88.

1904 BIRKHOFF, G.D. & VANDIVER, H.S.   On the integral divisors of $a^n - b^n$. Annals of Math., (2), 5, 1904, 173–180.

1904 CIPOLLA, M.   Sui numeri composti P, che verificano la congruenza di Fermat $a^{P-1} \equiv 1$ (mod P). Annali di Matematica, (3), 9, 1904, 139–160.

1912 CARMICHAEL, R.D.   On composite numbers P which satisfy the Fermat congruence $a^{P-1} \equiv 1$ (mod P). Amer. Math. Monthly, 19, 1912, 22–27.

1913 CARMICHAEL, R.D.   On the numerical factors of the arithmetic forms $\alpha^n \pm \beta^n$. Annals of Math., 15, 1913, 30–70.

1913 DICKSON, L.E.   Finiteness of odd perfect and primitive abundant numbers with n distinct prime factors. Amer. J. Math., 35, 1913, 413–422. Reprinted in *The Collected Mathematical Papers* (edited by A.A. Albert), Vol. I, 349–358. Chelsea, Bronx, N.Y., 1975.

1914 POCKLINGTON, H.C.   The determination of the prime or composite nature of large numbers by Fermat's theorem. Proc. Cambridge Phil. Soc., 18, 1914/6, 29–30.

1921 PIRANDELLO, L.   *Fu Mattia Pascal.* Bemporad & Figlio, Firenze, 1921, 169 (multiple editions and translations into many languages; especially recommended.)

1922 CARMICHAEL, R.D.   Note on Euler's ϕ-function. Bull. Amer. Math. Soc., 28, 1922, 109–110.

1925 CUNNINGHAM, A.J.C. & WOODALL, H.J.   *Factorization* of $y^n \pm 1$, $y = 2, 3, 5, 6, 7, 10, 11, 12$ *Up to High Powers(n)*. Hodgson, London, 1925.

1929 PILLAI, S.S.   On some functions connected with $\phi(n)$. Bull. Amer. Math. Soc., 35, 1929, 832–836.

1930 LEHMER, D.H.   An extended theory of Lucas' functions. Annals of Math., 31, 1930, 419–448. Reprinted in *Selected Papers* (edited by D. McCarthy), Vol. I, 11–48. Ch. Babbage Res. Centre, St. Pierre, Manitoba, Canada, 1981.

1932 LEHMER, D.H.   On Euler's totient function. Bull. Amer. Math. Soc., 38, 1932, 745–757. Reprinted in *Selected Papers* (edited by D. McCarthy), Vol. I, 319–325. Ch. Babbage Res. Centre, St. Pierre, Manitoba, Canada, 1981.

1932 WESTERN, A.E.   On Lucas' and Pepin's tests for the primeness of Mersenne's numbers. J. London Math. Soc., 7, 1932, 130–137.

1935 ARCHIBALD, R.C.   Mersenne's numbers. Scripta Math., 3, 1935, 112–119.

1935 LEHMER, D.H. On Lucas' test for the primality of Mersenne numbers. J. London Math. Soc., 10, 1935, 162–165. Reprinted in *Selected Papers* (edited by D. McCarthy), Vol. I, 86–89. Ch. Babbage Res. Centre, St. Pierre, Manitoba, Canada, 1981.

1936 LEHMER, D.H.   On the converse of Fermat's theorem. Amer. Math. Monthly, 43, 1936, Reprinted in *Selected Papers* (edited by D. McCarthy), Vol. I, 90–95. Ch. Babbage Res. Centre, St. Pierre, Manitoba, Canada, 1981.

1939 CHERNICK, J.   On Fermat's simple theorem. Bull. Amer. Math. Soc., 45, 1939, 269–274.

1944 PILLAI, S.S.   On the smallest primitive root of a prime. J. Indian Math. Soc., (N.S.), 8, 1944, 14–17.

1944 SCHUH, F.   Can $n - 1$ be divisible by $\phi(n)$ when n is composite? (in Dutch). Mathematica, Zutphen, B, 12, 1944, 102–107.

1945 KAPLANSKY, I.   Lucas' tests for Mersenne numbers. Amer. Math. Monthly, 52, 1945, 188–190.

1947 KLEE, V.L.   On a conjecture of Carmichael. Bull. Amer. Math. Soc., 53, 1947, 1183–1186.

1947 LEHMER, D.H.   On the factors of $2^n \pm 1$. Bull. Amer. Math. Soc., 53, 1947, 164–167. Reprinted in *Selected Papers* (edited by D. McCarthy), Vol. III, 1081–1084. Ch. Baggage Res. Centre, St. Pierre, Manitoba, Canada, 1981.

1948 ORE, O.   On the averages of the divisors of a number. Amer. Math. Monthly, 55, 1948, 615–619.

1949 ERDÖS, P.   On the convers of Fermat's theorem. Amer. Math. Monthly, 56, 1949, 623–624.

1949 FRIDLENDER, V.R.   On the least nth power non-residue (in Russian). Doklady Akad. Nauk SSSR (N.S.), 66, 1949, 351–352.

1949 SHAPIRO, H.N.   Note on a theorem of Dickson. Bull. Amer. Math. Soc., 55, 1949, 450–452.

1950 BEEGER, N.G.W.H.   On composite numbers n for which $a^{n-1} \equiv 1 \pmod{n}$, for every a, prime to n. Scripta Math., 16, 1950, 133–135.

1950 GIUGA, G.   Su una presumibile proprietà caratteristica dei numeri primi. Ist. Lombardo Sci. Lett. Rend. Cl. Sci. Mat. Nat., (3), 14 (83), 1950, 511–528.

1950 GUPTA, H.   On a problem of Erdös. Amer. Math. Monthly, 57, 1950, 326–329.

1950 KANOLD, H.J.   Sätze über Kreisteilungspolynome und ihre Anwendungen auf einige zahlentheoretische Probleme, II. Journal f.d. reine u. angew. Math., 187, 1950, 355–366.

1950 SALIÉ, H.   Über den kleinsten positiven quadratischen Nichtrest einer Primzahl. Math. Nachr., 3, 1949, 7–8.

1950 SOMAYAJULU, B.S.K.R.   On Euler's totient function $\phi(n)$. Math. Student, 18, 1950, 31–32.

1951 BEEGER, N.G.W.H.   On even numbers m dividing $2^m - 2$. Amer. Math. Monthly, 58, 1951, 553–555.

1952 DUPARC, H.J.A.   On Carmichael numbers. Simon Stevin, 29, 1952, 21–24.

1952 GRÜN, O.   Uber ungerade vollkommene Zahlen. Math. Zeits., 55, 1952, 353–354.

1954 KANOLD, H.J.   Über die Dichten der Mengen der vollkommenen und der befreundeten Zahlen. Math. Zeits., 61, 1954, 180–185.

1954 ROBINSON, R.M.   Mersenne and Fermat numbers. Proc. Amer. Math. Soc., 5, 1954, 842–846.

1954 SCHINZEL, A.   Quelques théorèmes sur les fonctions $\phi(n)$ et $\sigma(n)$. Bull. Acad. Polon. Sci., Cl. III, 2, 1954, 467–469.

1954 SCHINZEL, A.   Generalization of a theorem of B.S.K.R. Somayajulu on the Euler's function $\phi(n)$. Ganita, 5, 1954, 123–128.

1954 SCHINZEL, A. & SIERPIŃSKI, W.   Sur quelques propriétés des fonctions $\phi(n)$ et $\sigma(n)$. Bull. Acad. Polon. Sci., Cl. III, 2, 1954, 463–466.

1955 ARTIN, E.   The orders of the linear groups. Comm. Pure & Appl. Math., 8, 1955, 355–365. Reprinted in *Collected Papers* (edited by S. Lang & J.T. Tate), 387–397. Addison-Wesley, Reading, Mass., 1965.

1955 LABORDE, P.   A note on the even perfect numbers. Amer. Math. Monthly, 62, 1955, 348–349.

1956 SCHINZEL, A.   Sur l'équation $\phi(x) = m$. Elem. Math., 11, 1956, 75–78.

1956 SCHINZEL, A.   Sur un problème concernant la fonction $\phi(n)$. Czechoslovak Math. J., 6, (81), 1956, 164–165.

1958 JARDEN, D.   *Recurring Sequences.* Riveon Lematematika, Jerusalem, 1958 (3rd edition, Fibonacci Assoc., San Jose, CA, 1973).

1958 SIERPIŃSKI, W.   Sur les nombres premiers de la forme $n^n + 1$. L'Enseign. Math., (2), 4, 1958, 211–212.

1959 ROTKIEWICZ, A.   Sur les nombres pairs a pour lesquels les nombres $a^n b - ab^n$, respectivement $a^{n-1} - b^{n-1}$ sont divisibles par n. Rend. Circ. Mat. Palermo, (2), 8, 1959, 341–342.

1959 SCHINZEL, A.   Sur les nombres composés n qui divisent $a^n - a$. Rend. Circ. Mat. Palermo, (2), 7, 1958, 1–5.

1959 SCHINZEL, A. & WAKULICZ, A.   Sur l'équation $\phi(x+k) = \phi(x)$, II. Acta Arithm., 5, 1959, 425–426.

1959 WIRSING, E.   Bemerkung zu der Arbeit über vollkommene Zahlen. Math. Ann., 137, 1959, 316–318.

1960 INKERI, K.   Tests for primality. Annales Acad. Sci. Fennicae, Ser. A, I, 279, Helsinki, 1960, 19 pages.

1961 WARD, M.   The prime divisors of Fibonacci numbers. Pacific J. Math., 11, 1961, 379–389.

1962 BURGESS, D.A.   On character sums and L-series. Proc. London Math. Soc., (3), 12, 1962, 193–206.

1962 CROCKER, R.   A theorem on pseudo-primes. Amer. Math. Monthly, 69, 1962, p. 540.

1962 SCHINZEL, A.   The intrinsic divisors of Lehmer numbers in the case of negative discriminant. Arkiv. för Mat., 4, 1962, 413–416.

1962 SCHINZEL, A.   On primitive prime factors of $a^n - b^n$. Proc. Cambridge Phil. Soc., 58, 1962, 555–562.

1962 SHANKS, D.   *Solved and Unsolved Problems in Number Theory.* Spartan, Washington, 1962. 3rd edition by Chelsea, Bronx, N.Y., 1985.

1963 SCHINZEL, A.   On primitive prime factors of Lehmer numbers, I. Acta Arithm., 8, 1963, 211–223.

1963 SCHINZEL, A.   On primitive prime factors of Lehmer numbers, II. Acta Arithm., 8, 1963, 251–257.

1964 LEHMER, E.   On the infinitude of Fibonacci pseudo-primes. Fibonacci Quart., 2, 1964, 229–230.

1965 ROTKIEWICZ, A.   Sur les nombres de Mersenne dépourvus de facteurs carrés et sur les nombres naturels n tels que $n^2 | 2^n - 2$. Matem. Vesnik (Beograd), 2, (17), 1965, 78–80.

1966 GROSSWALD, E.   *Topics from the Theory of Numbers.* Macmillan, New York, 1966; 2nd edition Birkhäuser, Boston, 1984.

1966 MUSKAT, J.B.   On divisors of odd perfect numbers. Math. Comp., 20, 1966, 141–144.

1967 BRILLHART, J. & SELFRIDGE, J.L.   Some factorizations of $2^m \pm 1$ and related results. Math. Comp., 21, 1967, 87–96 and p. 751.

1967 MOZZOCHI, C.J.   A simple proof of the Chinese remainder theorem. Amer. Math. Monthly, 74, 1967, p. 998.

1970 LIEUWENS, E.   Do there exist composite numbers for which $k\phi(M) = M - 1$ holds? Nieuw. Arch. Wisk., (3), 18, 1970, 165–169.

1970 PARBERRY, E.A.   On primes and pseudo-primes related to the Fibonacci sequence. Fibonacci Quart., 8, 1970, 49–60.

1971 LIEUWENS, E.   *Fermat Pseudo-Primes.* Ph.D. Thesis, Delft, 1971.

1971 MORRISON, M.A. & BRILLHART, J.   The factorization of F_7. Bull. Amer. Math. Soc., 77, 1971, p. 264.

1972 HAGIS, Jr., P. & McDANIEL, W.L.   A new result concerning the structure of odd perfect numbers. Proc. Amer. Math. Soc., 32, 1972, 13–15.

1972 MILLS, W.H.   On a conjecture of Ore. Proc. 1972 Number Th. Conf. in Boulder, p. 142–146.

1972 RIBENBOIM, P.   *Algebraic Numbers.* Wiley-Interscience, New York, 1972.

1972 ROTKIEWICZ, A.   *Pseudoprime Numbers and their Generalizations.* Stud. Assoc. Fac. Sci. Univ. Novi Sad, 1972.

1973 ROTKIEWICZ, A.   On pseudoprimes with respect to the Lucas sequences. Bull. Acad. Polon. Sci., 21, 1973, 793–797.

1974 LIGH, S. & NEAL, L.   A note on Mersenne numbers. Math. Mag., 47, 1974, 231–233.

1974 POMERANCE, C.   On Carmichael's conjecture. Proc. Amer. Math. Soc., 43, 1974, 297–298.

1974 SCHINZEL, A.   Primitive divisors of the expression $A^n - B^n$ in algebraic number fields. Journal f. d. reine u. angew. Math., 268/269, 1974, 27–33.

1974 SINHA, T.N.   Note on perfect numbers. Math. Student, 42, 1974, p. 336.

1975 BRILLHART, J., LEHMER, D.H. & SELFRIDGE, J.L. New primality criteria and factorizations of $2^m \pm 1$. Math. Comp., 29, 1975, 620–647.

1975 GUY, R.K. How to factor a number. Proc. Fifth Manitoba Conf. Numerical Math., 1975, 49–89 (Congressus Numerantium, XVI, Winnipeg, Manitoba, 1976).

1975 HAGIS, Jr., P. & McDANIEL, W.L. On the largest prime divisor of an odd perfect number. Math. Comp., 29, 1975, 922–924.

1975 MORRISON, M.A. A note on primality testing using Lucas sequences. Math. Comp., 29, 1975, 181–182.

1975 POMERANCE, C. The second largest prime factor of an odd perfect number. Math. Comp., 29, 1975, 914–921.

1975 PRATT, V.R. Every prime has a succinct certificate. SIAM J. Comput., 4, 1975, 214–220.

1976 BUXTON, M. & ELMORE, S. An extension of lower bounds for odd perfect numbers. Notices Amer. Math. Soc., 23, 1976, p. A55.

1976 DIFFIE, W. & HELLMAN, M.E. New directions in cryptography. IEEE Trans. on Inf. Th., IT-22,

1976 LEHMER, D.H. Strong Carmichael numbers. J. Austral. Math. Soc., A, 21, 1976, 508–510.
Reprinted in *Selected Papers* (edited by D. McCarthy), Vol. I, 140–142. Ch. Babbage Res. Centre, St. Pierre, Manitoba, Canada, 1981.

1976 MENDELSOHN, N.S. The equation $\phi(x) = k$. Math. Mag., 49, 1976, 37–39.

1976 MILLER, G.L. Riemann's hypothesis and tests for primality. J. Comp. Syst. Sci., 13, 1976, 300–317.

1976 RABIN, M.O. Probabilistic algorithms. In *Algorithms and Complexity* (edited by J.F. Traub), 21–39. Academic Press, New York, 1976.

1976 YORINAGA, M. On a congruential property of Fibonacci numbers. Numerical experiments. Considerations and Remarks. Math. J. Okayama Univ., 19, 1976, 5–10; 11–17.

1977 MALM, D.E.G. On Monte-Carlo primality tests. Notices Amer. Math. Soc., 24, 1977, A-529, abstract 77T-A22.

1977 POMERANCE, C. On composite n for which $\phi(n)|n-1$. II Pacific J. Math., 69, 1977, 177–186.

1977 POMERANCE, C. Multiply perfect numbers, Mersenne

primes and effective computability. Math. Ann., 226, 1977, 195–206.

1977 SOLOVAY, R. & STRASSEN, V.   A fast Monte-Carlo test for primality. SIAM J. Comput., 6, 1977, 84–85.

1977 STEWART, C.L.   Primitive divisors of Lucas and Lehmer numbers. In *Transcendence Theory: Advances and Applications* (edited by A. Baker & D.W. Masser), 79–92. Academic Press, London, 1977.

1977 WILLIAMS, H.C.   On numbers analogous to the Carmichael numbers. Can. Math. Bull., 20, 1977, 133–143.

1978 KISS, P. & PHONG, B.M.   On a function concerning second order recurrences. Ann. Univ. Sci. Budapest, 21, 1978, 119–122.

1978 RIVEST, R.L.   Remarks on a proposed cryptanalytic attack on the M.I.T. public-key cryptosystem. Cryptologia, 2, 1978.

1978 RIVEST, R.L., SHAMIR, A. & ADLEMAN, L.M.   A method for obtaining digital signatures and public-key cryptosystems. Comm. ACM, 21, 1978, 120–126.

1978 WILLIAMS, H.C.   Primality testing on a computer. Ars Comb., 5, 1978, 127–185.

1978 YORINAGA, M.   Numerical computation of Carmichael numbers. Math. J. Okayama Univ., 20, 1978, 151–163.

1979 CHEIN, E.Z.   Non-existence of odd perfect numbers of the form $q_1^{a_1} q_2^{a_2} \cdots q_6^{a_6}$ and $5^{a_1} q_2^{a_2} \cdots q_9^{a_9}$. Ph.D. Thesis, Pennsylvania State Univ., 1979.

1980 BAILLIE, R. & WAGSTAFF, Jr., S.S.   Lucas pseudoprimes. Math. Comp., 35, 1980, 1391–1417.

1980 COHEN, G.L. & HAGIS, Jr., P.   On the number of prime factors of n if $\phi(n)|(n-1)$. Nieuw Arch. Wisk., (3), 28, 1980, 177–185.

1980 HAGIS, Jr., P.   Outline of a proof that every odd perfect number has at least eight prime factors. Math. Comp., 35, 1980, 1027–1032.

1980 MONIER, L.   Evaluation and comparison of two efficient probabilistic primality testing algorithms. Theoret. Comput. Sci., 12, 1980, 97–108.

1980 POMERANCE, C., SELFRIDGE, J.L. & WAGSTAFF, Jr., S.S.   The pseudoprimes to 25.10^9. Math. Comp., 35,

1980, 1003–1026.

1980 RABIN, M.O.   Probabilistic algorithm for testing primality. J. Nb. Th., 12, 1980, 128–138.

1980 WAGSTAFF, Jr., S.S.   Large Carmichael numbers. Math. J. Okayama Univ., 22, 1980, 33–41.

1980 WALL, D.W.   Conditions for $\phi(N)$ to properly divide $N-1$. In *A Collection of Manuscripts Related to the Fibonacci Sequence* (edited by V.E. Hoggatt & M. Bicknell-Johnson), 205–208. 18th Anniv. Vol., Fibonacci Assoc., San Jose, 1980.

1981 BRENT, R.P. & POLLARD, J.M.   Factorization of the eighth Fermat number. Math. Comp., 36, 1981, 627–630.

1981 GROSSWALD, E.   On Burgess' bound for primitive roots modulo primes and an application to $\Gamma(p)$. Amer. J. Math., 103, 1981, 1171–1183.

1981 LENSTRA, Jr., H.W.   Primality testing algorithms (after Adleman, Rumely and Williams). In Séminaire Bourbaki, exposé No. 576. Lecture Notes in Math., #901, 243-257. Springer-Verlag, Berlin, 1981.

1981 LÜNEBURG, H.   Ein einfacher Beweis für den Satz von Zsigmondy über primitive Primteiler von $A^N - 1$. In *Geometries and Groups* (edited by M. Aigner & D. Jungnickel). Lecture Notes in Math., #893, 219-222. Springer-Verlag, New York, 1981.

1982 BRENT, R.P.   Succinct proofs of primality for the factors of some Fermat numbers. Math. Comp., 38, 1982, 253–255.

1982 LENSTRA, Jr., H.W.   Primality testing. In *Computational Methods in Number Theory* (edited by H.W. Lenstra, Jr. & R. Tijdeman), Part I, 55–77. Math. Centre Tracts, #154, Amsterdam, 1982.

1982 MASAI, P. & VALETTE, A.   A lower bound for a counterexample in Carmichael's conjecture. Boll. Un. Mat. Ital., (6), 1-A, 1982, 313–316.

1982 WOODS, D. & HUENEMANN, J.   Larger Carmichael numbers. Comput. Math. & Appl., 8, 1982, 215–216.

1983 ADLEMAN, L.M., POMERANCE, C. & RUMELY, R.S.   On distinguishing prime numbers from composite numbers. Annals of Math., (2), 117, 1983, 173-206.

1983 BRILLHART, J., LEHMER, D.H., SELFRIDGE, J.L., TUCKERMAN, B. & WAGSTAFF, Jr., S.S.   *Factor-*

izations of $b^n \pm 1$, $b = 2$, 3, 5, 6, 7, 10, 11, 12 up to High Powers. Contemporary Math., Vol. 22, Amer. Math. Soc., Providence, R.I., 1983, second edition, 1988).

1983 HAGIS, Jr., P.   Sketch of a proof that an odd perfect number relatively prime to 3 has at least eleven prime factors. Math. Comp., 40, 1983, 399–404.

1983 POMERANCE, C. & WAGSTAFF, Jr., S.S.   Implementation of the continued fraction integer factoring algorithm. Congressus Numerantium, 37, 1983, 99–118.

1983 POWELL, B.   Problem 6420 (On primitive roots). Amer. Math. Monthly, 90, 1983, p. 60.

1983 SINGMASTER, D.   Some Lucas pseudoprimes. Abstracts Amer. Math. Soc., 4, 1983, No. 83T-10-146, p. 197.

1983 YATES, S.   Titanic primes. J. Recr. Math., 16, 1983/4, 250–260.

1984 COHEN, H. & LENSTRA, Jr., H.W.   Primality testing and Jacobi sums. Math. Comp., 42, 1984, 297–330.

1984 DIXON, J.D.   Factorization and primality tests. Amer. Math. Monthly, 91, 1984, 333–352.

1984 KEARNES, K.   Solution of problem 6420. Amer. Math. Monthly, 91, 1984, p. 521.

1984 NICOLAS, J.L.   Tests de primalité. Expo. Math., 2, 1984, 223–234.

1984 POMERANCE, C.   *Lecture Notes on Primality Testing and Factoring* (Notes by G.M. Gagola Jr.). Math. Assoc. America, Notes, No. 4, 1984, 34 pages.

1984 WILLIAMS, H.C.   An overview of factoring. In *Advances in Cryptology* (edited by D. Chaum), 71–80. Plenum, New York, 1984.

1984 YATES, S.   Sinkers of the Titanics. J. Recr. Math., 17, 1984/5, 268–274.

1985 BEDOCCHI, E.   Note on a conjecture on prime numbers. Rev. Mat. Univ. Parma (4), 11, 1985, 229–236.

1985 RIESEL, H.   *Prime Numbers and Computer Methods for Factorization.* Birkhäuser, Boston, 1985.

1985 WAGON, S.   Perfect numbers. Math. Intelligencer, 7, No. 2, 1985, 66–68.

1986 KISS, P., PHONG, B.M., & LIEUWENS, E.   On Lucas pseudoprimes which are products of s primes. In *Fibonacci*

Numbers and their Applications edited by A.N. Philippou, G.E. Bergum & A.F. Horadam), 131–139. Reidel, Dordrecht, 1986.

1986 WAGON, S.   Carmichael's "Empirical Theorem". Math. Intelligencer, 8, No. 2, 1986, 61–62.

1986 WAGON, S.   Primality testing. Math. Intelligencer, 8, No. 3, 1986, 58–61.

1987 COHEN, G.L.   On the largest component of an odd perfect number. J. Austral. Math. Soc. (Ser. A), 42, 1987, 280–286.

1987 KOBLITZ, N.   *A Course in Number Theory and Cryptography.* Springer-Verlag, New York, 1987.

1987 LENSTRA, Jr., H.W.   Factoring integers with elliptic curves. Annals Math, 126, 1987, 649–673.

1988 BRILLHART, J., MONTGOMERY, P.L. & SILVER-MAN, R.D.   Tables of Fibonacci and Lucas factorizations, and Supplement. Math. Comp., 50, 1988, 251–260 and S-1 to S-15.

1988 YOUNG, J. & BUELL, D.A.   The twentieth Fermat number is composite. Math. Comp., 50, 1988, 261–262.

1989 BATEMAN, P.T., SELFRIDGE, J.L., & WAGSTAFF, Jr., S.S. The New Mersenne Conjecture, Amer. Math. Monthly, 96, 1989, 125–128.

1989 BRENT, R.A. & COHEN, G.L.   A new lower bound for odd perfect numbers. Math. Comp., 53, 1989, 431–437 and S-7 to S-24.

1989 DUBNER, H. A new method for producing large Carmichael numbers. Math. Comp., 53, 1989, 411–414.

1989 LEMOS, M.   *Criptografia, Números Primos e Algoritmos.* (17° Colóquio Brasileiro de Matemática) Inst. Mat. Pura e Aplic., Rio de Janeiro, 1989, 72 pages.

1989 LENSTRA, A.K. & LENSTRA, Jr., H.W.   Algorithms in number theory, in Handbook of Theoretical Computer Science, North-Holland, Amsterdam, 1989.

1990 LENSTRA, A.K., LENSTRA, Jr., H.W., MANASSE, M.S. & POLLARD, J.M. The number theory sieve. Preprint, 1990.

Chapter 3

1912 FROBENIUS, F.G.   Über quadratische Formen, die viele Primzahlen darstellen. Sitzungsber. d. Königl. Akad. d. Wiss. zu

Berlin, 1912, 966-980. Reprinted in *Gesammelte Abhandlungen*, Vol. III, 573–587. Springer-Verlag, Berlin, 1968.

1912 RABINOVITCH, G. Eindeutigkeit der Zerlegung in Primzahlfaktoren in quadratischen Zahlkörpern. Proc. Fifth Intern. Congress Math., Cambridge, Vol. 1, 1912, 418–421.

1938 SKOLEM, T. *Diophantische Gleichungen.* Springer-Verlag, Berlin, 1938.

1947 MILLS, W.H. A prime-representing function. Bull. Amer. Math. Soc., 53, p. 604.

1951 WRIGHT, E.M. A prime-representing function. Amer. Math. Monthly, 58, 1951, 616–618.

1952 HEEGNER, K. Diophantische Analysis und Modulfunktionen. Math. Zeits., 56, 1952, 227–253.

1960 PUTNAM, H. An unsolvable problem in number theory. J. Symb. Logic, 1960, 220–232.

1962 COHN, H. *Advanced Number Theory.* Wiley, New York, 1962. Reprinted by Dover, New York, 1980.

1964 WILLANS, C.P. On formulae for the nth prime. Math. Gaz., 48, 1964, 413–415.

1966 BAKER, A. Linear forms in the logarithms of algebraic numbers. Mathematika, 13, 1966, 204–216.

1967 STARK, H.M. A complete determination of the complex quadratic fields of class-number one. Michigan Math. J., 14, 1967, 1–27.

1969 DUDLEY, U. History of a formula for primes. Amer. Math. Monthly, 76, 1969, 23–28.

1971 GANDHI, J.M. Formulae for the nth prime. Proc. Washington State Univ. Conf. on Number Theory, 96–106. Wash. St. Univ., Pullman, Wash., 1971.

1971 MATIJASEVIČ, Yu.V. Diophantine representation of the set of prime numbers (in Russian). Dokl. Akad. Nauk SSSR, 196, 1971, 770–773. English translation by R. N. Goss, in Soviet Math. Dokl. 11, 1970, 354–358.

1972 VANDEN EYNDEN, C. A proof of Gandhi's formula for the nth prime. Amer. Math. Monthly, 79, 1972, p. 625.

1973 DAVIS, M. Hilbert's tenth problem is unsolvable. Amer. Math. Monthly, 80, 1973, 233–269.

1973 KARST, E. New quadratic forms with high density of primes. Elem. d. Math., 28, 1973, 116–118.

1974 GOLOMB, S.W.   A direct interpetation of Gandhi's formula. Amer. Math. Monthly, 81, 1974, 752–754.

1974 HENDY, M.D.   Prime quadratics associated with complex quadratic fields of class number two. Proc. Amer. Math. Soc., 43, 1974, 253–260.

1975 ERNVALL, R.   A formula for the least prime greater than a given integer. Elem. d. Math., 30, 1975, 13–14.

1975 JONES, J.P.   Diophantine representation of the Fibonacci numbers. Fibonacci Quart., 13, 1975, 84–88.

1975 MATIJASEVIČ, Yu.V.   Reduction of an arbitrary diophantine equation to one in 13 unknowns. Acta Arithm., 27, 1975, 521–553.

1976 JONES, J.P., SATO, D., WADA, H. & WIENS, D. Diophantine representation of the set of prime numbers. Amer. Math. Monthly, 83, 1976, 449–464.

1977 MATIJASEVIČ, Yu.V.   Primes are nonnegative values of a polynomial in 10 variables. Zapiski Sem. Leningrad Mat. Inst. Steklov, 68, 1977, 62–82. English translation by L. Guy & J.P. Jones, J. Soviet Math., 15, 1981, 33–44.

1979 JONES, J.P.   Diophantine representation of Mersenne and Fermat primes, Act. Arith., 35, 1979, 209–221.

1988 RIBENBOIM, P.   Euler's famous prime generating polynomial and the class number of imaginary quadratic fields. L'Enseign. Math., 34, 1988, 23–42.

1989 GOETGHELUCK, P.   On cubic polynomials giving many primes. Elem. d. Math., 44, 1989, 70–73.

Chapter 4

1885 MEISSEL, E.D.F.   Berechnung der Menge von Primzahlen, welche innerhalb der ersten Milliarde natürlicher Zahlen vorkommen. Math. Ann., 25, 1885, 251–257.

1892 SYLVESTER, J.J.   On arithmetical series. Messenger of Math., 21, 1892, 1–19 and 87–120. Reprinted in *Gesammelte Abhandlungen*, Vol. III, 573–587. Springer-Verlag, New York, 1968.

1901 VON KOCH, H.   Sur la distribution des nombres premiers. Acta Math., 24, 1901, 159–182.

1901 WOLFSKEHL, P.   Ueber eine Aufgabe der elementaren Arithmetik. Math. Ann., 54, 1901, 503–504.

1909 LANDAU, E.   *Handbuch der Lehre von der Verteilung der Primzahlen.* Teubner, Leipzig, 1909. Reprinted by Chelsea, Bronx, N.Y., 1974.

1914 LITTLEWOOD, J.E.   Sur la distribution des nombres premiers. C.R. Acad. Sci. Paris, 158, 1914, 869–872.

1919 BRUN, V.   Le crible d'Eratosthène et le théorème de Goldbach. C.R. Acad. Sci. Paris, 168, 1919, 544–546.

1919 BRUN, V.   La série $\frac{1}{5} + \frac{1}{7} + \frac{1}{11} + \frac{1}{13} + \frac{1}{17} + \frac{1}{19} + \frac{1}{29} + \frac{1}{31} + \frac{1}{41} + \frac{1}{43} + \frac{1}{59} + \frac{1}{61} + \cdots$ où les dénominateurs sont "nombres premiers jumeaux" est convergente ou finie. Bull. Sci. Math., (2), 43, 1919, 100–104 and 124–128.

1920 BRUN, V.   Le crible d'Erathostène et la théorème de Goldbach. Videnskapsselskapets Skrifter Kristiania, Mat.-nat. Kl. 1920, No. 3, 36 pages.

1923 HARDY, G. H. & LITTLEWOOD, J.E.   Some problems of "Partitio Numerorum", III: On the expression of a number as a sum of primes. Acta Math., 44, 1923, 1–70. Reprinted in *Collected Papers of G.H. Hardy*, Vol. I, 561–630. Clarendon Press, Oxford, 1966.

1930 ERDÖS, P.   Beweis eines Satzes von Tschebycheff. Acta Sci. Math. Szeged, 5, 1930, 194–198.

1930 HOHEISEL, G.   Primzahlprobleme in der Analysis. Sitzungsberichte Berliner Akad. d. Wiss., 1930, 580–588.

1930 SCHNIRELMANN, L.   Über additive Eigenschaften von Zahlen. Ann. Inst. Polytechn. Novocerkask, 14, 1930, 3–28 and Math. Ann., 107, 1933, 649–690.

1933 SKEWES, S.   On the difference $\pi(x) - li(x)$. J. London Math. Soc., 8, 1933, 277–283.

1934 ISHIKAWA, H.   Uber die Verteilung der Primzahlen. Sci. Rep. Tokyo Bunrika Daigaku, A, 2, 1934, 27–40.

1937 CRAMÉR, H.   On the order of magnitude of the difference between consecutive prime numbers. Acta Arithm., 2, 1937, 23–46.

1937 INGHAM, A.E.   On the difference between consecutive primes. Quart. J. Pure & Appl. Math., Oxford, Ser. 2, 8, 1937, 255–266.

1937 LANDAU, E.   Über einige neuere Fortschritte der additiven Zahlentheorie. Cambridge Univ. Press, Cambridge, 1937. Reprinted by Stechert-Hafner, New York, 1964.

1937 VAN DER CORPUT, J.G.   Sur l'hypothèse de Goldbach pour presque tous les nombres pairs. Acta Arith., 2, 1937, 266–290.

1937 VINOGRADOV, I.M.   Representation of an odd number as the sum of three primes (in Russian). Dokl. Akad. Nauk SSSR, 15, 1937, 169–172.

1938 ESTERMANN, T.   Proof that almost all even positive integers are sums of two primes. Proc. London Math. Soc., 44, 1938, 307–314.

1938 POULET, P.   Table des nombres composés vérifiant le théorème de Fermat pour le module 2, jusqu' à 100.000.000. Sphinx, 8, 1938, 52-52 Corrections: Math. Comp., 25, 1971, 944–945 and 26, 1972, p. 814.

1938 ROSSER, J.B.   The nth prime is greater than $n \log n$. Proc. London Math. Soc. 45, 1938, 21–44.

1938 TSCHUDAKOFF, N.G.   On the density of the set of even integers which are not representable as a sum of two odd primes (in Russian). Izv. Akad. Nauk SSSR, Ser. Mat., 1, 1938, 25–40.

1939 VAN DER CORPUT, J.G.   Über Summen von Primzahlen und Prim- zahlquadraten. Math. Ann., 116, 1939, 1–50.

1944 LINNIK, Yu.V.   On the least prime in an arithmetic progression I. The basic theorem (in Russian). Mat. Sbornik, 15 (57), 1944, 139–178.

1946 BRAUER, A.   On the exact number of primes below a given limit. Amer. Math. Monthly, 9, 1946, 521–523.

1947 KHINCHIN, A.Ya.   *Three Pearls of Number Theory.* Original Russian edition in OGIZ, Moscow, 1947. Translation into English published by Graylock Press, Baltimore, 1952.

1947 RÉNYI, A.   On the representation of even numbers as the sum of a prime and an almost prime. Dokl. Akad. Nauk SSSR, 56, 1947, 455–458.

1949 CLEMENT, P.A.   Congruences for sets of primes. Amer. Math. Monthly, 56, 1949, 23–25.

1949 ERDÖS, P.   On a new method in elementary number theory which leads to an elementary proof of the prime number theorem. Proc. Nat. Acad. Sci. U.S.A., 35, 1949, 374–384.

1949 MOSER, L.   A theorem on the distribution of primes.

Amer. Math. Monthly, 56, 1949, 624–625.

1949 RICHERT, H.E.   Über Zerfällungen in ungleiche Primzahlen. Math. Zeits., 52, 1949, 342–343.

1949 SELBERG, A.   An elementary proof of the prime number
theorem. Annals of Math., 50, 1949, 305–313.

1949 SELBERG, A.   An elementary proof of Dirichlet's theorem
about primes in an arithmetic progression. Annals of Math., 50,
1949, 297–304.

1949 SELBERG, A.   An elementary proof of the prime number
theorem for arithmetic progressions. Can. J. Math., 2, 1950,
66–78.

1950 ERDÖS, P.   On almost primes. Amer. Math. Monthly, 57,
1950, 404–407.

1950 HASSE, H.   *Vorlesungen über Zahlentheorie.* Springer-Verlag,
Berlin, 1950.

1950 SELBERG, A.   The general sieve method and its place
in prime number theory. Proc. Int. Congr. Math., Cambridge,
1950.

1951 TITCHMARSH, E.C.   *The Theory of the Riemann Zeta
Function.* Clarendon Press, Oxford, 1951.

1956 ERDÖS, P.   On pseudo-primes and Carmichael numbers.
Publ. Math. Debrecen, 4, 1956, 201–206.

1957 LEECH, J.   Note on the distribution of prime numbers. J.
London Math. Soc., 32, 1957, 56–58.

1958 SCHINZEL, A. & SIERPIŃSKI, W.   Sur certaines hypothèses concernant les nombres premiers. Acta Arithm., 4,
1958, 185–208; Erratum, 5, 1959, p. 259.

1959 SCHINZEL, A.   Démonstration d'une conséquence de l'hypothèse de Goldbach. Compositio Math., 14, 1959, 74–76.

1961 WRENCH, J.W.   Evaluation of Artin's constant and the
twin-prime constant. Math. Comp., 15, 1961, 396–398.

1962 ROSSER, J.B. & SCHOENFELD, L.   Approximate formulas for some functions of prime numbers. Illinois J. Math.,
6, 1962, 64–94.

1963 AYOUB, R.G.   *An Introduction to the Theory of Numbers.*
Amer. Math. Soc., Providence, R.I., 1963.

1963 KANOLD, H.J.   Elementare Betrachtungen zur Primzahl-
theorie. Arch. Math., 14, 1963, 147–151.

1963 ROTKIEWICZ, A.   Sur les nombres pseudo-premiers de la forme $ax + b$. C.R. Acad. Sci. Paris, 257, 1963, 2601–2604.

1963 WALFISZ, A.Z. *Weylsche Exponentialsummen in der neueren Zahlentheorie.* VEB Deutscher Verlag d. Wiss., Berlin, 1963.

1965 GELFOND, A.O. & LINNIK, Yu.V. *Elementary Methods in Analytic Number Theory.* Translated by A. Feinstein, revised and edited by L.J. Mordell. Rand McNally, Chicago, 1965.

1965 PAN, C.D. On the least prime in an arithmetic progression. Sci. Record (N.S.), 1, 1957, 311–313.

1965 ROTKIEWICZ, A. Les intervalles contenant les nombres pseudo premiers. Rend. Circ. Mat. Palermo (2), 14, 1965, 278–280.

1965 STEIN, M.L. & STEIN, P.R. New experimental results on the Goldbach conjecture. Math. Mag., 38, 1965, 72–80.

1965 STEIN, M.L. & STEIN, P.R. Experimental results on additive 2-bases. Math. Comp., 19, 1965, 427–434.

1966 BOMBIERI, E. & DAVENPORT, H. Small differences between prime numbers. Proc. Roy. Soc., A, 293, 1966, 1–18.

1967 LANDER, L.J. & PARKIN, T.R. Consecutive primes in arithmetic progression. Math. Comp., 21, 1967, p. 489.

1967 ROTKIEWICZ, A. On the pseudo-primes of the form $ax + b$. Proc. Cambridge Phil. Soc., 63, 1967, 389–392.

1967 SZYMICZEK, K. On pseudo-primes which are products of distinct primes. Amer. Math. Monthly, 74, 1967, 35–37.

1969 MONTGOMERY, H.L. Zeros of L-functions. Invent. Math., 8, 1969, 346–354.

1969 ROSSER, J.B., YOHE, J.M. & SCHOENFELD, L. Rigorous computation of the zeros of the Riemann zeta-function (with discussion). Inform. Processing 68 (Proc. IFIP Congress, Edinburgh, 1968), Vol. I, 70–76. North-Holland, Amsterdam, 1969.

1971 TITCHMARSH, E.C. *The Theory of the Riemann Zeta Function.* Clarendon Press, Oxford, 1951.

1972 HUXLEY, M.N. On the difference between consecutive primes. Invent. Math., 15, 1972, 164–170.

1972 HUXLEY, M.N. *The Distribution of Prime Numbers.* Oxford Univ. Press, Oxford, 1972.

1972 ROTKIEWICZ, A. On a problem of W. Sierpiński. Elem. d. Math., 27, 1972, 83–85.

1973/1978 CHEN, J.R.   On the representation of a large even integer as the sum of a prime and the product of at most two primes, I and II. Sci. Sinica, 16, 1973, 157–176; and 21, 1978, 421–430.

1973 MONTGOMERY, H.L.   The pair correlation of zeros of the zeta function. *Analytic Number Theory* (Proc. Symp. Pure Math., Vol. XXIV, St. Louis, 1972), 181–193. Amer. Math. Soc., Providence, R.I., 1973.

1974 AYPUB, R.B.   Euler and the zeta-function. Amer. Math. Monthly, 81, 1974, 1067–1086.

1974 EDWARDS, H.M.   *Riemann's Zeta Function.* Academic Press, New York, 1974.

1974 HALBERSTAM, H. & RICHERT, H.E.   *Sieve Methods.* Academic Press, New York, 1974.

1974 LEVINSON, N.   More than one third of zeros of Riemann's zeta function are on $\sigma = 1/2$. Adv. in Math., 13, 1984, 383–436.

1974 MAKOWSKI, A.   On a problem of Rotkiewicz on pseudo-primes. Elem. d. Math., 29, 1974, p. 13.

1975 MONTGOMERY, H.L. & VAUGHAN, R.C.   The exceptional set in Goldbach's problem. Acta Arithm., 27, 1975, 353–370.

1975 ROSS, P.M.   On Chen's theorem that each large even number has the form $p_1 + p_2$ or $p_1 + p_2 p_3$. J. London Math. Soc., (2), 10, 1975, 500–506.

1975 ROSSER, J.B. & SCHOENFELD, L.   Sharper bounds for Chebyshev functions $\theta(x)$ and $\psi(x)$. Math. Comp., 29, 1975, 243–269.

1976 APOSTOL, T.M.   *Introduction to Analytic Number Theory.* Springer-Verlag, New York, 1976.

1976 BRENT, R.P.   Tables concerning irregularities in the distribution of primes and twin primes to 10^{11}. Math. Comp., 30, 1976, p. 379.

1977 HUDSON, R.H.   A formula for the exact number of primes below a given bound in any arithmetic progression. Bull. Austral. Math. Soc., 16, 1977, 67-73.

1977 HUDSON, R.H. & BRAUER, A.   On the exact number of primes in the arithmetic progressions $4n \pm 1$ and $6n \pm 1$. Journal f. d. reine u. angew. Math., 291, 1977, 23–29.

1977 LANGEVIN, M. Méthodes élémentaires en vue du théorème de Sylvester. Sém. Delange-Pisot-Poitou, 17e année, 1975/76, fasc. 1, exp. No. G12, 9 pages, Paris, 1977.

1977 WEINTRAUB, S.   Seventeen primes in arithmetic progression. Math. Comp., 31, 1977, p. 1030.

1978 BAYS, C. & HUDSON, R.H.   On the fluctuations of Littlewood for primes of the form $4n \pm 1$. Math. Comp., 32, 141, 281–286.

1978 HEATH-BROWN, D.R.   Almost-primes in arithmetic progressions and short intervals. Math. Proc. Cambridge Phil. Soc., 83, 1978, 357–375.

1979 HEATH-BROWN, D.R. & IWANIEC, H.   On the difference between consecutive powers. Bull. Amer. Math. Soc., N.S., 1, 1979, 758–760.

1979 IWANIEC, H. & JUTILA, M.   Primes in short intervals. Arkiv f. Mat., 17, 1979, 167–176.

1979 POMERANCE, C.   The prime number graph. Math. Comp., 33, 1979, 399–408.

1979 WAGSTAFF, Jr., S.S.   Greatest of the least primes in arithmetic progressions having a given modulus. Math. Comp., 33, 1979, 1073–1080.

1980 CHEN, J.R. & PAN, C.D.   The exceptional set of Goldbach numbers, I. Sci. Sinica, 23, 1980, 416–430.

1980 LIGHT, W.A., FORREST, J., HAMMOND, N., & ROE, S.   A note on Goldbach's conjecture. BIT, 20, 1980, p. 525.

1980 NEWMAN, D.J.   Simple analytic proof of the prime number theorem. Amer. Math. Monthly, 87, 1980, 693–696.

1980 PINTZ, J.   On Legendre's prime number formula. Amer. Math. Monthly, 87, 1980, 733–735.

1980 POMERANCE, C., SELFRIDGE, J.L., & WAGSTAFF, Jr., S.S.   The pseudoprimes to $25 \cdot 10^9$. Math. Comp., 35, 1980, 1003–1026.

1980 VAN DER POORTEN, A.J. & ROTKIEWICZ, A.   On strong pseudoprimes in arithmetic progressions. J. Austral. Math. Soc., A, 29, 1980, 316–321.

1981 HEATH-BROWN, D.R.   Three primes and an almost prime in arithmetic progression. J. London Math. Soc., (2), 23, 1981, 396–414.

1981 POMERANCE, C.   On the distribution of pseudo-primes.

Math. Comp., 37, 1981, 587–593.

1982 POMERANCE, C.   A new lower bound for the pseudo-primes counting function. Illinois J. Math., 26, 1982, 4–9.

1982 WEINTRAUB, S.   A prime gap of 682 and a prime arithmetic sequence. BIT, 22, 1982, p. 538.

1983 CHEN, J.R.   The exceptional value of Goldbach numbers, II. Sci. Sinica, Ser. A, 26, 1983, 714–731.

1983 POWELL, B.   Problem 6429 (Difference between consecutive primes). Amer. Math. Monthly, 90, 1983, p. 338.

1983 RIESEL, H. & VAUGHAN, R.C.   On sums of primes. Arkiv f. Mat., 21, 1983, 45–74.

1984 DAVIES, R.O.   Solution of problem 6429. Amer. Math. Monthly, 91, 1984, p. 64.

1984 IWANIEC, H. & PINTZ, J.   Primes in short intervals. Monatsh. Math., 98, 1984, 115–143.

1984 SCHROEDER, M.R.   *Number Theory in Science and Communication.* Springer-Verlag, New York, 1984.

1984 WANG, Y.   *Goldbach Conjecture.* World Scientific Publ., Singapore, 1984.

1985 IVIĆ, A.   *The Riemann Zeta-Function.* J. Wiley & Sons, New York, 1985.

1985 LAGARIAS, J.C., MILLER, V.S. & ODLYZKO, A.M.   Computing $\pi(x)$: The Meissel-Lehmer method. Math. Comp., 44, 1985, 537-560.

1985 LOU, S. & YAO, Q.   The upper bound of the difference between consecutive primes. Kexue Tongbao, 8, 1985, 128–129.

1985 POWELL, B.   Problem 1207 (A generalized weakened Goldbach theorem). Math. Mag., 58, 1985, p. 46; 59, 1986, 48–49.

1985 PRITCHARD, P.A.   Long arithmetic progressions of primes; some old, some new. Math. Comp., 45, 1985, 263–267.

1986 BOMBIERI, E., FRIEDLANDER, J.B. & IWANIEC, H.   Primes in arithmetic progression to large moduli, I. Acta Math., 156, 1986, 203–251.

1986 FINN, M.V. & FROHLIGER, J.A.   Solution of problem 1207. Math. Mag., 59, 1986, 48–49.

1986 MOZZOCHI, C.J.   On the difference between consecutive primes. J. Nb. Th., 24, 1986, 181–187.

1986 TE RIELE, H.J.J.   On the sign of the difference $\pi(x) - \ell i(x)$. Report NM-R8609, Centre for Math. and Comp. Science,

Amsterdam, 1986; Math. Comp., 48, 1987, 323–328.

1986 VAN DE LUNE, J., TE RIELE, H.J.J., & WINTER, D.T.   On the zeros of the Riemann zeta function in the critical strip, IV. Math. Comp., 47, 1986, 667–681.

1986 WAGON, S.   Where are the zeros of zeta of s? Math. Intelligencer, 8, 4, 1986, 57–62.

1988 ERDÖS, P., KISS, P. & SÁRKÖZY, A.   A lower bound for the counting function of Lucas pseudoprimes. Math. Comp., 51, 1988, 315–323.

1988 PATTERSON, S.J.   *Introduction to the Theory of the Riemann Zeta-function.* Cambridge Univ. Press, Cambridge, 1988.

1989 CONREY, J.B.   At least two fifths of the zeros of the Riemann zeta function are on the critical line. Bull. Amer. Math. Soc., 20, 1989, 79–81.

1989 YOUNG, J. & POTLER, A.   First occurrence of prime gaps. Math. Comp., 52, 1989, 221–224.

1990 JAESCHKE, G.   The Carmichael numbers to 10^{12}. Math. Comp., 55, 1990, 383–389.

1990 PARADY, B.K., SMITH, J.F. & ZARANTONELLO, S.   Largest known twin primes. Math. Comp., 55, 1990, 381–382.

Chapter 5

1948 GUNDERSON, N.G.   *Derivation of Criteria for the First Case of Fermat's Last Theorem and the Combination of these Criteria to Produce a New Lower Bound for the Exponent.* Ph.D. Thesis, Cornell University, 1948, 111 pages.

1951 DÉNES, P.   An extension of Legendre's criterion in connection with the first case of Fermat's last theorem. Publ. Math. Debrecen, 2, 1951, 115–120.

1953 GOLDBERG, K.   A table of Wilson quotients and the third Wilson prime. J. London Math. Soc., 28, 1953, 252–256.

1954 WARD, M.   Prime divisors of second order recurring sequences. Duke Math. J., 21, 1954, 607–614.

1956 OBLÁTH, R.   Une propriété des puissances parfaites. Mathesis, 65, 1956, 356–364.

1960 SIERPIŃSKI, W.   Sur un problème concernant les nombres $k \cdot 2^n + 1$. Elem. d. Math., 15, 1960, 73–74.

1963 BRILLHART, J.   Some miscellaneous factorizations. Math. Comp., 17, 1963, 447–450.

1964 SIEGEL, C.L.   Zu zwei Bemerkungen Kummers. Nachr. Akad. d. Wiss. Göttingen, Math. Phys. Kl., II, 1964, 51–62. Reprinted in *Gesammelte Abhandlungen* (edited by K. Chandrasckharan & M. Maaß), Vol. III, 436–442. Springer-Verlag, Berlin, 1966.

1965 KLOSS, K.E.   Some number theoretic calculations. J. Res. Nat. Bureau of Stand., B, 69, 1965, 335–336.

1966 HASSE, H.   Uber die Dichte der Primzahlen p, fur die eine vorgegebene ganzrationale Zahl $a \neq 0$ von gerader bzw. ungerader Ordnung mod p ist. Math. Ann., 168, 1966, 19–23.

1966 KRUYSWIJK, D.   On the congruence $u^{p-1} \equiv 1 \pmod{p^2}$ (in Dutch). Math. Centrum Amsterdam, 1966, 7 pages.

1971 BRILLHART, J., TONASCIA, J., & WEINBERGER, P.J.   On the Fermat quotient. *Computers in Number Theory* (edited by A.L. Atkin & B.J. Birch), 213–222. Academic Press, New York, 1971.

1975 JOHNSON, W.   Irregular primes and cyclotomic invariants. Math. Comp., 29, 1975, 113–120.

1978 WAGSTAFF, Jr., S.S.   The irregular primes to 125000. Math. Comp., 32, 1978, 583–591.

1978 WILLIAMS, H.C.   Some primes with interesting digit patterns. Math. Comp., 32, 1978, 1306–1310.

1979 ERDÖS, P. & ODLYZKO, A.M.   On the density of odd integers of the form $(p-1)2^{-n}$ and related questions. J. Nb. Th., 11, 1979, 257–263.

1979 RIBENBOIM, P.   *13 Lectures on Fermat's Last Theorem.* Springer-Verlag, New York, 1979.

1979 WILLIAMS, H.C. & SEAH, E.   Some primes of the form $(a^n - 1)/(a - 1)$. Math. Comp., 33, 1979, 1337–1342.

1980 POWELL, B.   Primitive densities of certain sets of primes. J. Nb. Th., 12, 1980, 210–217.

1981 LEHMER, D.H.   On Fermat's quotient, base two. Math. Comp., 36, 1981, 289–290.

1982 POWELL, B.   Problem E 2956 (The existence of small prime solutions of $x^{p-1} \not\equiv 1 \pmod{p^2}$). Amer. Math. Monthly, 89, 1982, p. 498.

1982 YATES, S.   *Repunits and Repetends.* Star Publ. Co., Boynton Beach, Florida, 1982.

1983 JAESCHKE, G.   On the smallest k such that $k \cdot 2^N + 1$ are composite. Corrigendum. Math. Comp., 40, 1983, 381-384; 45, 1985, p. 637.

1983 KELLER, W.   Factors of Fermat numbers and large primes of the form $k \cdot 2^n + 1$. Math. Comp., 41, 1983, 661–673.

1983 RIBENBOIM, P.   1093. Math. Intelligencer, 5, No. 2, 1983, 28–34.

1985 LAGARIAS, J.C.   The set of primes dividing the Lucas numbers has density $\frac{2}{3}$. Pacific J. Math., 118, 1985, 19–23.

1986 TZANAKIS, N.   Solution to problem E 2956. Amer. Math. Monthly, 93, 1986, p. 569.

1986 WILLIAMS, H.C. & DUBNER, H.   The primality of R1031. Math. Comp., 47, 1986, 703–712.

1987 GRANVILLE, A.J.   *Diophantine Equations with Variable Exponents with Special Reference to Fermat's Last Theorem.* Ph.D. Thesis, Queen's University, Kingston, 1987, 207 pages.

1987 ROTKIEWICZ, A.   Note on the diophantine equation $1 + x + x^2 + \cdots + x^n = y^m$. Elem. d. Math., 42, 1987, p. 76.

1987 TANNER, J.W. & WAGSTAFF, Jr., S.S.   New congruences for the Bernoulli numbers. Math. Comp., 48, 1987, 341–350.

1988 GONTER, R.H. & KUNDERT, E.G.   Wilson's theorem (mod p) (n-1)! $\equiv$ -1 (modp^2), has been computed up to 10,000,000. Fourth SIAM Conference on Discrete Mathematics, San Francisco, June 19, 1988.

1988 GRANVILLE, A. & MONAGAN, M.B.   The first case of Fermat's last theorem is true for all prime exponents up to 714, 591, 416, 091 389. Trans. Amer. Math. Soc., 306, 1988, 329–359.

1989 LÖH, G.   Long chains of nearly doubled primes. Math. Comp., 53, 1989, 751–759.

1989 TANNER, J.W. & WAGSTAFF, Jr., S.S.   New bound for the first case of Fermat's last theorem. Math. Comp., 53, 1989, 743–750.

1990 COPPERSMITH, D.   Fermat's last theorem (case I) and the Wieterich criterion. Math. Comp., 54, 1990, 895–902.

1991 BUHLER, J.P., CRANDALL, R.E., & SOMPOLSKI, R.W.   Irregular primes to one million (preprint).

1991 FEE, G. & GRANVILLE, A.   The prime factors of Wendt's binomial circulant determinant. To appear.

1991 SUZUKI, J.   Some computations of the generalized Wieferich criteria. Math. Comp., 56, 1991 (to appear).

Chapter 6

1857 BOUNIAKOWSKY, V.   Nouveaux théorèmes relatifs à la distribution des nombres premiers et à la décomposition des entiers en facteurs. Mém. Acad. Sci. St. Petersbourg, (6), Sci. Math. Phys., 6, 1857, 305–329.

1904 DICKSON, L.E.   A new extension of Dirichlet's theorem on prime numbers. Messenger of Math., 33, 1904, 155–161.

1922 NAGELL, T.   Zur Arithmetik der Polynome. Abhandl. Math. Sem. Univ. Hamburg, 1, 1922, 179–194.

1923 HARDY, G.H. & LITTLEWOOD, J.E.   Some problems in "Partitio Numerorum", III: On the expression of a number as a sum of primes. Acta Math., 44, 1923, 1–70. Reprinted in *Collected Papers of G.H. Hardy*, Vol. I, 561–630. Clarendon Press, Oxford, 1966.

1931 HEILBRONN, H.   Über die Verteilung der Primzahlen in Polynomen. Math. Ann., 104, 1931, 794–799.

1958 SCHINZEL, A. & SIERPIŃSKI, W.   Sur certaines hypothèses concernant les nombres premiers. Remarque. Acta Arithm., 4, 1958, 185–208 and 5, 1959, p. 259.

1961 SCHINZEL, A.   Remarks on the paper "Sur certaines hypothèses concernant les nombres premiers". Acta Arithm., 7, 1961, 1–8.

1964 SIERPIŃSKI, W.   Les binômes $x^2 + n$ et les nombres premiers. Bull. Soc. Roy. Soc. Liège, 33, 1964, 259–260.

1969 RIEGER, G.J.   On polynomials and almost-primes. Bull. Amer. Math. Soc., 75, 1969, 100–103.

1974 HALBERSTAM, H. & RICHERT, H.E.   *Sieve Methods*. Academic Press, New York, 1974.

1975 SHANKS, D.   Calculation and applications of Epstein zeta functions. Math. Comp. 29, 1975, 271–287.

1978 IWANIEC, H.   Almost-primes represented by quadratic polynomials. Invent. Math., 47, 1978, 171–188.

1981 POWELL, B.   Problem 6384 (Numbers of the form $m^p -$

n). Amer. Math. Monthly, 89, 1982, p. 278.

1982 ISRAEL, R.B.   Solution of problem 6384. Amer. Math. Monthly, 90, 1983, p. 650.

1984 McCURLEY, K.S.   Prime values of polynomials and irreducibility testing. Bull. Amer. Math. Soc., 11, 1984, 155–158.

1986 McCURLEY, K.S.   The smallest prime value of $x^n + a$. Can. J. Math., 38, 1986, 925–936.

1986 McCURLEY, K.S.   Polynomials with no small prime values. Proc. Amer. Math. Soc., 97, 1986, 393–395.

1990 FUNG, G.W. & WILLIAMS, H.C.   Quadratic polynomials which have a high density of prime values. Math. Comp., 55, 1990, 345-353.

Primes up to 10,000

2	3	5	7	11	13	17	19	23	29	31	37	41	43	47	53	59	61	67	71	73	79	83	89
97	101	103	107	109	113	127	131	137	139	149	151	157	163	167	173	179	181	191	193	197	199	211	223
227	229	233	239	241	251	257	263	269	271	277	281	283	293	307	311	313	317	331	337	347	349	353	359
367	373	379	383	389	397	401	409	419	421	431	433	439	443	449	457	461	463	467	479	487	491	499	503
509	521	523	541	547	557	563	569	571	577	587	593	599	601	607	613	617	619	631	641	643	647	653	659
661	673	677	683	691	701	709	719	727	733	739	743	751	757	761	769	773	787	797	809	811	821	823	827
829	839	853	857	859	863	877	881	883	887	907	911	919	929	937	941	947	953	967	971	977	983	991	997
1009	1013	1019	1021	1031	1033	1039	1049	1051	1061	1063	1069	1087	1091	1093	1097	1103	1109	1117	1123	1129	1151	1153	1163
1171	1181	1187	1193	1201	1213	1217	1223	1229	1231	1237	1249	1259	1277	1279	1283	1289	1291	1297	1301	1303	1307	1319	1321
1327	1361	1367	1373	1381	1399	1409	1423	1427	1429	1433	1439	1447	1451	1453	1459	1471	1481	1483	1487	1489	1493	1499	1511
1523	1531	1543	1549	1553	1559	1567	1571	1579	1583	1597	1601	1607	1609	1613	1619	1621	1627	1637	1657	1663	1667	1669	1693
1697	1699	1709	1721	1723	1733	1741	1747	1753	1759	1777	1783	1787	1789	1801	1811	1823	1831	1847	1861	1867	1871	1873	1877
1879	1889	1901	1907	1913	1931	1933	1949	1951	1973	1979	1987	1993	1997	1999	2003	2011	2017	2027	2029	2039	2053	2063	2069
2081	2083	2087	2089	2099	2111	2113	2129	2131	2137	2141	2143	2153	2161	2179	2203	2207	2213	2221	2237	2239	2243	2251	2267
2269	2273	2281	2287	2293	2297	2309	2311	2333	2339	2341	2347	2351	2357	2371	2377	2381	2383	2389	2393	2399	2411	2417	2423
2437	2441	2447	2459	2467	2473	2477	2503	2521	2531	2539	2543	2549	2551	2557	2579	2591	2593	2609	2617	2621	2633	2647	2657
2659	2663	2671	2677	2683	2687	2689	2693	2699	2707	2711	2713	2719	2729	2731	2741	2749	2753	2767	2777	2789	2791	2797	2801
2803	2819	2833	2837	2843	2851	2857	2861	2879	2887	2897	2903	2909	2917	2927	2939	2953	2957	2963	2969	2971	2999	3001	3011
3019	3023	3037	3041	3049	3061	3067	3079	3083	3089	3109	3119	3121	3137	3163	3167	3169	3181	3187	3191	3203	3209	3217	3221
3229	3251	3253	3257	3259	3271	3299	3301	3307	3313	3319	3323	3329	3331	3343	3347	3359	3361	3371	3373	3389	3391	3407	3413
3433	3449	3457	3461	3463	3467	3469	3491	3499	3511	3517	3527	3529	3533	3539	3541	3547	3557	3559	3571	3581	3583	3593	3607
3613	3617	3623	3631	3637	3643	3659	3671	3673	3677	3691	3697	3701	3709	3719	3727	3733	3739	3761	3767	3769	3779	3793	3797
3803	3821	3823	3833	3847	3851	3853	3863	3877	3881	3889	3907	3911	3917	3919	3923	3929	3931	3943	3947	3967	3989	4001	4003
4007	4013	4019	4021	4027	4049	4051	4057	4073	4079	4091	4093	4099	4111	4127	4129	4133	4139	4153	4157	4159	4177	4201	4211
4217	4219	4229	4231	4241	4243	4253	4259	4261	4271	4273	4283	4289	4297	4327	4337	4339	4349	4357	4363	4373	4391	4397	4409

4421	4423	4441	4447	4451	4457	4463	4481	4483	4493	4507	4513	4517	4519	4523	4547	4549	4561	4567	4583	4591	4597	4603	4621
4637	4639	4643	4649	4651	4657	4663	4673	4679	4691	4703	4721	4723	4729	4733	4751	4759	4783	4787	4789	4793	4799	4801	4813
4817	4831	4861	4871	4877	4889	4903	4909	4919	4931	4933	4937	4943	4951	4957	4967	4969	4973	4987	4993	4999	5003	5009	5011
5021	5023	5039	5051	5059	5077	5081	5087	5099	5101	5107	5113	5119	5147	5153	5167	5171	5179	5189	5197	5209	5227	5231	5233
5237	5261	5273	5279	5281	5297	5303	5309	5323	5333	5347	5351	5381	5387	5393	5399	5407	5413	5417	5419	5431	5437	5441	5443
5449	5471	5477	5479	5483	5501	5503	5507	5519	5521	5527	5531	5557	5563	5569	5573	5581	5591	5623	5639	5641	5647	5651	5653
5657	5659	5669	5683	5689	5693	5701	5711	5717	5737	5741	5743	5749	5779	5783	5791	5801	5807	5813	5821	5827	5839	5843	5849
5851	5857	5861	5867	5869	5879	5881	5897	5903	5923	5927	5939	5953	5981	5987	6007	6011	6029	6037	6043	6047	6053	6067	6073
6079	6089	6091	6101	6113	6121	6131	6133	6143	6151	6163	6173	6197	6199	6203	6211	6217	6221	6229	6247	6257	6263	6269	6271
6277	6287	6299	6301	6311	6317	6323	6329	6337	6343	6353	6359	6361	6367	6373	6379	6389	6397	6421	6427	6449	6451	6469	6473
6481	6491	6521	6529	6547	6551	6553	6563	6569	6571	6577	6581	6599	6607	6619	6637	6653	6659	6661	6673	6679	6689	6691	6701
6703	6709	6719	6733	6737	6761	6763	6779	6781	6791	6793	6803	6823	6827	6829	6833	6841	6857	6863	6869	6871	6883	6899	6907
6911	6917	6947	6949	6959	6961	6967	6971	6977	6983	6991	6997	7001	7013	7019	7027	7039	7043	7057	7069	7079	7103	7109	7121
7127	7129	7151	7159	7177	7187	7193	7207	7211	7213	7219	7229	7237	7243	7247	7253	7283	7297	7307	7309	7321	7331	7333	7349
7351	7369	7393	7411	7417	7433	7451	7457	7459	7477	7481	7487	7489	7499	7507	7517	7523	7529	7537	7541	7547	7549	7559	7561
7573	7577	7583	7589	7591	7603	7607	7621	7639	7643	7649	7669	7673	7681	7687	7691	7699	7703	7717	7723	7727	7741	7753	7757
7759	7789	7793	7817	7823	7829	7841	7853	7867	7873	7877	7879	7883	7901	7907	7919	7927	7933	7937	7949	7951	7963	7993	8009
8011	8017	8039	8053	8059	8069	8081	8087	8089	8093	8101	8111	8117	8123	8147	8161	8167	8171	8179	8191	8209	8219	8221	8231
8233	8237	8243	8263	8269	8273	8287	8291	8293	8297	8311	8317	8329	8353	8363	8369	8377	8387	8389	8419	8423	8429	8431	8443
8447	8461	8467	8501	8513	8521	8527	8537	8539	8543	8563	8573	8581	8597	8599	8609	8623	8627	8629	8641	8647	8663	8669	8677
8681	8689	8693	8699	8707	8713	8719	8731	8737	8741	8747	8753	8761	8779	8783	8803	8807	8819	8821	8831	8837	8839	8849	8861
8863	8867	8887	8893	8923	8929	8933	8941	8951	8963	8969	8971	8999	9001	9007	9011	9013	9029	9041	9043	9049	9059	9067	9091
9103	9109	9127	9133	9137	9151	9157	9161	9173	9181	9187	9199	9203	9209	9221	9227	9239	9241	9257	9277	9281	9283	9293	9311
9319	9323	9337	9341	9343	9349	9371	9377	9391	9397	9403	9413	9419	9421	9431	9433	9437	9439	9461	9463	9467	9473	9479	9491
9497	9511	9521	9533	9539	9547	9551	9587	9601	9613	9619	9623	9629	9631	9643	9649	9661	9677	9679	9689	9697	9719	9721	9733
9739	9743	9749	9767	9769	9781	9787	9791	9803	9811	9817	9829	9833	9839	9851	9857	9859	9871	9883	9887	9901	9907	9923	9929
9931	9941	9949	9967	9973	10007	10009	10037	10039	10061	10067	10069	10079	10091	10093									

Index of Tables

Index of Records

Index of Names

Gallimawfries

Subject Index